WINGS AND THE NAVY

WINGS AND THE NAVY

COLIN JONES

Kangaroo Press

Cover picture: A Sea Fury based at Lee-on-the-Solent taking off from the *Sydney*, 1949.

© Colin Jones 1997

First published in 1997 by Kangaroo Press Pty Ltd
3 Whitehall Road, Kenthurst NSW 2156 Australia
P.O. Box 6125 Dural Delivery Centre NSW 2158
Printed by Australian Print Group, Maryborough 3465

ISBN 0 86417 836 0

Contents

Introduction 7

1. Building a New Navy 9
2. Showing the Flag 18
3. The End of the Beginning 26
4. His Majesty's Australian Fleet 35
5. Anzac Brothers 45
6. The Shadow of the New War 52
7. The Life of the Ship 62
8. The Test of Battle 73
9. The Way Forward 94

Afterword 107
Glossary 109
Appendices 111
Bibliography 116
Notes 119
Index 121

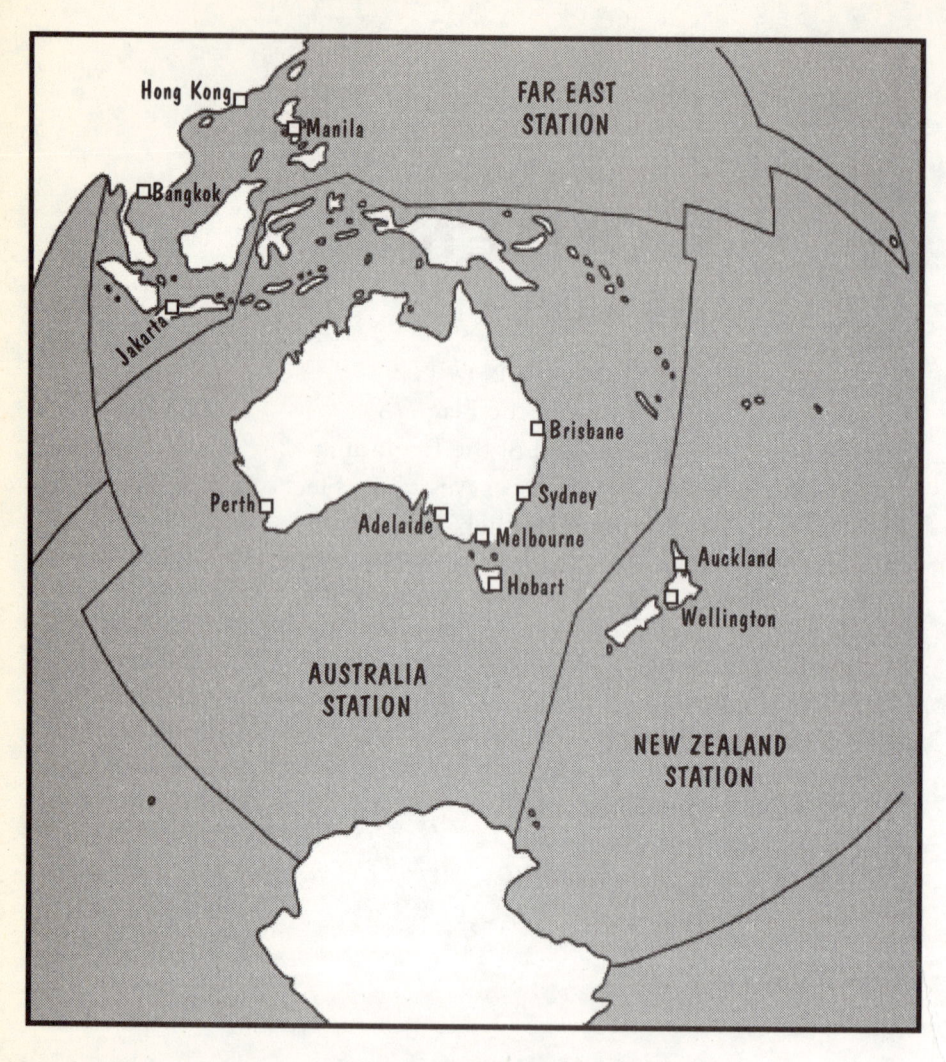

NAVAL STATIONS IN THE PACIFIC OCEAN.
(ORTHOGRAPHIC PROJECTION BASED ON MELBOURNE)

INTRODUCTION

The Royal Australian Navy was founded in 1911 in the form which it was to follow for the next 35 years, as a squadron of cruisers and a flotilla of destroyers, the sort of force which would fit into a scheme of defence appropriate for the far-flung empire of Great Britain. It gained its first laurels early, in the defeat of the German cruiser *Emden* by the *Sydney* at Cocos Island in 1914. In the Second World War the men of the navy performed with dedication and high professional skill in all theatres of action, but their ships were rarely the most modern and the inevitable attrition of battle left their ranks sadly depleted when peace came in 1945.

Many histories have been written of the Second World War, of tactics and strategy, of experience and heroism. The navy was always to the forefront. It took heavy losses, and acquitted itself in the finest traditions of the service. Nevertheless, as the war came to an end, the Australian navy was a much reduced force, lacking in that essential of a modern fleet, an aircraft carrier. If the navy was to continue and grow after the war, the government of the time would have to actively consider the most appropriate path of development. This book is the story of those plans and ideals, and how they were fulfilled.

I have endeavoured in this work to remedy in part the absence of a detailed postwar naval history, especially in what might be called 'the arts of peace'. I have therefore not chosen to commence on the day when the guns fell silent, for that exact date is open to some dispute. The last naval bombardment of the Second World War by an Australian ship was by the *Gascoyne* on 26 July 1945. On the other hand, British troops, ships and aircraft were still in action in November as part of the re-establishment of Dutch rule in its Indonesian possessions, and right through to October, Australian ships were present for the surrender of isolated Japanese garrisons. They had to repatriate troops, prisoners and internees, to clear mines and obstructions, and to maintain order in a world where order had been shaken to its very foundations. Foreign bases remained on the Australian mainland until early 1946.

Thus, a time has to be left for restoring, to some degree, the normal rhythms of life before commencing the story. This book, then, is a story of the postwar period, when big things were expected of the Australian navy, and before cutbacks modified the vision of the future. It is in particular the story of the arrival of the aircraft carrier in our navy, along with the associated provisions of the 1947 five-year plan.

The fleet war games of the period provide an insight into the way that the operations of the new navy were conceived, and they contrast nicely with the experience of HMAS *Sydney* in the Korean War.

While far from being a comprehensive history of the navy from 1947 to 1953, this story attempts to fill a gap, and to revive the hopes and the experiences of the postwar world. It is easy to impose the perspective of a later generation on the times, and argue that things should have been done differently, but I have endeavoured to resist this, and instead present my story in its contemporary context.

It is appropriate to begin on 28 May 1947, the day on which the ensign and pennant were again raised aboard the pride of the navy, the heavy cruiser *Australia*. After strenuous service and significant war damage, she had been extensively refitted and now emerged a new ship. She displayed the latest wireless, radar, fighter direction and communications installations, as well as improved accommodation and crew amenities.

With her return to service, the *Shropshire* was put into reserve, but the *Hobart* continued to share Australian commitments in the occupation of Japan. Also involved with occupation duties were two units of the Tenth Destroyer Flotilla and the troopship *Kanimbla*. Apart from necessary duties in clearing away the remains and the dislocation of the war, there was a major need for the navy to show the flag in adjacent seas and lend visible proof to the victory over Japan.

The First Frigate Flotilla provided vessels for service in New Guinea, while the Twentieth Minesweeping Flotilla, comprising a sloop, six corvettes, a stores lighter and four motor launches, was sweeping the Barrier Reef passages from Townsville north. They had to find and destroy some 4654 mines which had closed the area to shipping for six years.

Two other frigates and a sloop were engaged in survey work and the *Manoora* was trooping.

The navy was far from idle, but what was its role? Was it no more than a detached squadron of the Royal Navy? The answer came in a major government policy statement.

1

BUILDING A NEW NAVY

On 4 June 1947 the Australian Minister for Defence in the Chifley Labor Government, John Dedman, announced a large program of naval re-equipment over the next five years.[1] It was part of a major policy revision for the three services and the highest peacetime defence vote in Australian history. In the first year, it was expected that expenditure would run to:

RAN	£15 million
Army	£12.5 million
RAAF	£12.5 million
Dept of Munitions	£10 million, including £6 million for research and the guided weapons project.[2]

It was realised that during the war the navy was far from a complete or balanced force. Its cruisers had been important but subsidiary units and the major victories in the Pacific had been won solely by aircraft carriers. The image of the relentless pounding of the Japanese home islands by aircraft from no less than 20 U.S. and British aircraft carriers in July and August 1945 emphasised the point. Canada had manned two Royal Navy escort carriers from 1944, and was proceeding with the creation of a permanent carrier force. Indeed, it seemed that a fleet without an aircraft carrier would no longer be worthy of the name.

For this reason the largest allocation of the new program was for the provision of two fleet aircraft carriers. These would be complemented by the six destroyers already on order, of which two, *Anzac* and *Tobruk*, were under construction. As for the cruiser squadron — depleted by battle and with its most modern unit 11 years old — the obvious superiority of the new destroyers, especially the big Daring class, left the future open to speculation. The new destroyers' 4.5 inch guns were capable of a rate of fire of up to 25 rounds per minute. Some saw them as a new generation of light cruisers.★

★ It is ironic that when the Germans first saw the V and W class destroyers in the Great War they thought they were encountering light cruisers. By the Second World War they were able to deride these same vessels as so much 'scrap iron'.

A comprehensive building program was important according to John Dedman: 'It is essential that a navy must be ready to fight as soon as a war occurs. The main vessels, which take a long time to build, must be in existence before war occurs,' the Minister said.[3]

He then enunciated the idea that had haunted the Australian navy since its creation in 1911, the idea of an Australian fleet that could stand on its own.

> The naval programme aims at building up a balanced force over a period of years, which will be capable of operating as an independent force, backed by shore establishments for its maintenance. It also includes escort vessels for the protection of our shipping and survey vessels to continue the surveys necessary in Australian waters.[4]

At a cost of £75 000 000, Australia would have a strong and compact striking force, capable of operating independently as a task force, or in co-operation with Allied naval forces, and it was noted that the strength of the navy would be twice that of 1939. By 1953 it was planned to have the following composition:

Squadron	2	light fleet carriers
	2	cruisers
	6	destroyers
Escort Force	3	frigates
Surveying Duties	3	survey ships
	3	survey tenders
Training	1	frigate
	2	minesweepers
	3	air/sea rescue vessels
Auxiliary Vessels	1	ocean-going tug
	1	ammunition vessel
	2	boom defence vessels
79 vessels in reserve including:		
	1	cruiser
	2	destroyers
	6	frigates
	31	minesweepers[5]

Base facilities at Sydney, including the new £9 000 000 Captain Cook Dock, completed early in 1945 and capable of taking the largest capital ships, were regarded with some considerable pride. They had been vital for the wartime operation of the British Pacific Fleet. The base at Garden

Island was soon to receive a new crane, with a lifting capacity of 250 tons. In addition the base at Lombrun Point, Manus Island, was to be developed as the major base for the Australian navy in northern waters. This would replace HMAS *Tarangau*, the establishment at Dreger Harbour, on the Huon Peninsula, New Guinea. The government, it was stated, 'would welcome an arrangement for its use by the U.S.A. on the principle of reciprocity'.[6] Indeed, it was hoped that if the U.S. Navy found the base useful it might also contribute capital to its development. The naval air station at Nowra, near Jervis Bay, would be developed for the Australian navy, to provide a shore base for the new carrier-borne air squadrons, and a second air station would be provided at Schofields, near Sydney. Jervis Bay had deep water suitable for aircraft carriers, and land and sea bombing ranges well away from built-up areas.

During the war the air stations had been part of the web of bases that supported the British Pacific Fleet and its 17 'flat-tops'. Nowra had been HMS *Nabbington* until October 1945, but instead of being closed, it received HMS *Nabswick* from Jervis Bay until April 1946. Schofields had been HMS *Nabthorpe* until November 1945, when HMS *Nabstock* was transferred south from Maryborough in Queensland until May 1946. It was one of the last of the British Pacific Fleet facilities in Australia to close.

The development of a shipbuilding industry was also an important part of Australian defence policy. For this reason, the option of purchasing destroyers from Britain had been rejected. The British had been surprised when they came to survey Australian maritime infrastructure in the latter part of the war, that resources were so thin. This should not be the case a second time.

RAN manpower was to be substantially increased, on a staged basis, reaching 14 753 by 1952, a rise in the vicinity of 50 per cent. Naval aviation, on consideration, was to be a part of the navy, and not the air force. Thus Australia was not to repeat the error of Canada where naval aircraft were operated by the air force, and did not become part of the navy until September 1948. There had, indeed, been some consideration in Australia as to whether the RAAF should remain in its present form, or whether it should be split with the other services. The end result was to a great extent the work of Group Captain Valston Hancock, the RAAF representative on the planning team, who favoured the option of a force structure similar to that of Britain, in which the navy would control its own aircraft.

New 12-year-service recruiting resumed in September 1945, with 2537 signing up for the navy in the first six months. At the start of demobilisation, in October 1945, the strength of the navy stood at 39 000, but fell steadily thereafter as people returned to civilian life. More than 28 000 people were released from the navy by the end of September 1946. There was to be no revival of compulsory military training. The legislation was not repealed, but it was allowed to lie dormant.

Mr Dedman said that the program gave 'very practical and substantial effect to the Prime Minister's promise to undertake a greater share of the burden of British Commonwealth defence'. An illustration of this was that in war all Empire navies would 'act under strategical direction of the British Admiralty'.[7] An aviation component would therefore be very significant in assisting British forces — already facing severe restraint in the postwar financial environment. At least there was not a single prospective enemy with a surface fleet worth mentioning. The signal failure of Britain to provide adequate naval resources in the East Indies and the Far East between 1941 and 1944 was, however, remembered. There would be no more reliance on fortress Singapore, and critics pointed to Australia's open flank to the north west, where in the main, the guard of the Royal Navy had ceased to exist. On the British side, it was intended that Australia should be a more significant element in the defence structure in the region than it had been in the past. On the basis of building up a balanced and realistic force structure east of Suez, the Commonwealth forces would notionally comprise squadrons based around two aircraft carriers and two cruisers in the Far East and also in Australia, while New Zealand and the East Indies would have forces whose strongest units would be two cruisers. Like most postwar schemes, this was to be little more than ideas on a piece of paper, but it seemed achievable at the time.

The Americans were in strength in the Pacific, and in occupation in Japan, while the Dutch and French also had substantial forces in the region retaking power in their colonies. In March 1946, the French cruiser *Emile Bertin* bombarded Chinese troops at Haiphong in support of a landing, and during 1947 the escort carrier *Dixmude* operated her Dauntless dive bombers over Vietnam. Dutch Fireflies were in action against Indonesian nationalists in the same year.

For all this, 'The Government realises that this program is a considerable insurance premium to pay as the price of security. But it is a small one when compared with the human and material costs of war. To rest content on a hard-won victory and let our defences run down, would be inviting a future catastrophe.'[8] The Australian fleet was not designed to fit into a preconceived force structure, despite the strong British connection. There could be no long-term reliance on a strong British fleet in the Far East, as bitter memory recalled. Our fleet had to be strong on its own. There was still a fear of the possible power of a resurgent Japan, and the Australian government saw it as far better for Australia and the British Commonwealth if forces were decentralised, with each member country concentrating on the defence of its own region. The official statement read:

> The Forces to be placed at the disposal of the United Nations for the maintenance of international peace and security, including regional arrangements in the Pacific.
>
> The Forces to be maintained under arrangements for co-operation in British Commonwealth Defence.
>
> The Forces to be maintained to provide for the inherent right of individual self-defence.[9]

While the completion of these objectives would be vigorously pursued, the policy would be kept flexible and a continuous review maintained in case the need for variation should arise. Whitehall approved the Australian decision:

> United Kingdom Government welcomes decision of Australian Government to form a naval aviation branch as part of its postwar defence policy and modernisation of Royal Australian Navy. Establishment of Australian naval air arm will be a valuable contribution to the naval resources of the British Commonwealth in the Pacific area.[10]

One of the main driving forces behind the push for Australian aircraft carriers was the First Naval Member, Vice-Admiral Sir Louis Hamilton, who had come from a distinguished wartime career, most recently in command at Malta. There had been expectations that Australia would commission the carrier *Ocean* and two modern cruisers in July 1945, but for a variety of reasons, this had not happened. Nevertheless, the navy was taking active steps in the latter part of 1945 to persuade the government to allow it to form an Air Branch.

Australia began official negotiations with Britain in April 1946 concerning the possible purchase of some of the ten incomplete light fleet aircraft carriers which could not be completed for, or manned by, the British service. The government was prepared to take a leisurely path, but Hamilton was determined to push matters along. Cabinet ministers and defence chiefs boarded the *Glory* off Flinders Naval Depot during her Australian cruise that year to watch a flying display. There had been talk of scrapping the uncompleted ships where they lay, but in order to ensure that they went to a force with close ties to the Royal Navy, Britain offered in September 1946 to pay £2 750 000, a figure which amounted to half the cost of the vessels. Outfits of armaments and stores would amount to a further £450 000 for each vessel.

From the British perspective, there was another cause to be served by making the offer. As Australia had been showing some disturbing signs of independence at the end of the war, it was appropriate that inducements should be provided to bring her back more firmly into the Commonwealth orbit. The aircraft carrier deal would be part of it.

The new naval program was modelled on classical approaches to sea power, as summarised by Admiral Sir Herbert Richmond in his 1946 book *Statesmen and Sea Power*. First, he said, a navy must have a strength element — its warships. Second, it must have a security element — bases which were properly defended. Third, it must have a transport element — a healthy merchant navy and domestic facilities for shipbuilding and repair.[11] He also thought that a navy would usually find itself on the strategic defensive and the tactical offensive at the start of a war. Its operations would be governed by the initial moves and advances of the enemy, but within these constraints, it should be able to strike out as and when required, to maximum effect. The concept would apply as equally to Australia as to Britain and it would be hoped that the new forces proposed would enable such strikes to be strong.

Apart from policy, preparations were under way in Australia for the hundred and one organisational details that were required to form a naval air arm. It was going to be a substantial job. Indeed, contemporary thinking in Russia held that the operation of aircraft carriers was such a difficult matter that its own navy should steer clear of them. Those in Russia who wanted carriers included in the postwar build-up of forces were to be constantly overruled. It might have been the same for Australia had we started from scratch, but we were able to draw on a strong vein of tradition and experience from the Royal Navy as it had evolved through the Second World War.

The light fleet carriers had been conceived in 1941, when new construction was urgently needed to replace wartime losses. The result was not a converted cruiser or passenger liner, but a carefully conceived warship.

The light fleet carriers, although small and relatively slow, were economical to operate, as they only had half the complement of a fleet carrier. Their lifts and hangars would accommodate the new larger aircraft. The flight decks were only 10 per cent shorter than those of fleet carriers, but they had six knots less speed. Their construction was a mixture of naval and mercantile practice, built up to the main deck under Lloyds' survey, and with many features to ensure simplicity of design. The engines, based on standard cruiser and destroyer equipment, were arranged in echelon in case of battle damage, and the compartmentation was such as to allow the ship to settle evenly if hit. This last detail was essential if aircraft were to be saved in a final emergency. Light anti-aircraft weapons only were carried. Detailed design work had been the responsibility of Vickers Armstrong, and the result was a very handsome vessel and one which, for its size, was to be remarkably versatile and durable.

The first keels were laid down in June 1942 and the last in October 1943. With much urgency surrounding the need for new carriers as the war progressed, the first, *Colossus*, was completed in December 1944. Sixteen of the class were proposed, but the last six were modified with stronger flight decks, larger lifts, better insulation, and improved accommodation. These, especially, were the ones which were offered for sale.

Two hulls were earmarked for Australian service on 3 June 1947 — the *Terrible* and *Majestic*. At the time when work was suspended in May 1946 their construction was about 80 per cent complete. The air squadrons for the first, and further advanced, ship were to comprise 12 Sea Fury single seat fighter-bombers, 12 Firefly two-seat anti-submarine reconnaissance and strike aircraft, and a Sea Otter amphibian for air/sea rescue. During the war, ships of the type had operated between 39 and 42 aircraft, but new aircraft were larger and heavier, and operational needs were no longer as critical.

Naval aircraft needed to be of a type to counter high-performance land-based types such as the Soviet Yak-3 fighter. The Hawker Sea Fury, which entered service in August 1947 as a naval version of an aircraft first proposed for the RAF, was possibly the best piston-engined fighter ever built. Mike Crosley, who was later to fly one from the *Ocean* in Korea, called it 'superb'.[12] Designed as a long-range fighter for the war in the Pacific, it was the last in a line of development from the Typhoon and Tempest. It was noted for its impressive acceleration and the lightness of its controls. The Fairey Firefly was slightly older, having seen its first action against the *Tirpitz* in 1944. Four squadrons served with the British Pacific Fleet. It was a good aircraft for the navy — rugged, dependable and docile, a low landing speed for deck landing, but of relatively high performance. It was said of the Firefly, 'What other than affection could be felt for a pilot's aeroplane that combined good handling characteristics with a sound performance, tremendous reliability and outstanding versatility.'[13]

The second Australian ship was to be completed to a modified design to take a specialised anti-submarine aircraft which eventually emerged as the

Gannet, and a jet fighter of limited performance, eventually the Sea Venom. The modifications, involving the strengthening of the deck and other features, were estimated to cost £500 000.

The capabilities of the new aircraft carriers were somewhat overstated by the British, who said that they would be 'capable of operating all the naval aircraft that will be in service in the middle '50s of this century and possibly well beyond that date'.[14] Indeed, it was anticipated that the ships might carry the new Wyvern strike aircraft and Sea Hawk jet fighters after 1951. In fact, the Wyvern suffered severe developmental problems and was not in service until 1953, the year the Sea Hawk was introduced. Neither was ever a realistic contender for Australian service, and the development of the Sea Venom went forward specifically because of the Australian need for an aircraft of this type. In 1948 the British naval staff agreed that, with the increased requirements of future aircraft, catapult launches would be the only real option. The *Sydney* had the latest model hydraulic catapult, 151 feet long, capable of launching an aircraft of 20 000 pounds weight at 66 knots. When operations were running smoothly, aircraft could land on her deck every 25 seconds.

Among the dominion navies, Canada was ahead of Australia in establishing naval aviation, as it had been acknowledged by expert opinion as early as 1945 that no modern navy, however small, should operate without an air component. The *Warrior* was commissioned in 1946 for the RCN, carrying 803 Squadron of Seafires, and 825 of Fireflies. In contrast to the Australian requirement, which saw the ships modified for tropical service, the *Warrior* was not sufficiently 'winterised' for Canada, and had to be withdrawn on a seasonal basis from the Atlantic to the milder climate of the Pacific. She was replaced by the *Magnificent* in 1948. As with the Australian carriers, Canada received a group of warships at a substantial discount, though in her case it had been accepted that the defence budget could not support two carriers.

Along with the carriers for Australia went the new destroyers. One of these, the *Tobruk*, was launched at Cockatoo Island on 20 December 1947 and she boasted the most powerful marine engines ever built in Australia.

Following the Australian decision, the British First Aircraft Carrier Squadron visited to showcase naval aviation with a public demonstration of its operations, and to engage in combined manoeuvres with the Australian fleet. In the major exercise, the British fleet, sailing from Melbourne, would be 'attacked' by heavy bombers near Gabo and Montagu Islands, and would then be engaged by the Australian fleet, sailing from Sydney.

The Lincoln heavy bomber was new in the RAAF at this time. After delays in the supply of the engines from England, the first had flown on 17

March 1946 and aircraft were working up at East Sale before being moved to their permanent squadron location at Amberley. Of the two visiting carriers, the *Glory* had previously visited Australia in 1945 as part of the British Pacific Fleet, and had been at the scene of the Japanese surrender at Rabaul. After being used on repatriation duties at the end of the year, she re-embarked her aircraft, and visited ports including Adelaide and Newcastle in mid-1946. Since then she operated with both the British Pacific Fleet and the East Indies Fleet. The other carrier, *Theseus*, was a new ship, little over a year from completion.

2

SHOWING THE FLAG

The *Theseus* and *Cockade* arrived in Hobart from Singapore on 6 July 1947 in a bitter squall of rain. They berthed at Station Pier in Melbourne on 11 July and there they were joined by the *Glory* and *Contest*, which had experienced a wild stormy voyage from Adelaide. As they came up Port Phillip in 40 knot winds, seven Seafires and six Fireflies took off from *Theseus*, in 2 minutes 30 seconds, for a flight over the city. One Firefly remained at the rear end of the deck with engine trouble when the crew of a Seafire signalled that they wished to make an emergency landing because of trouble with a wing flap. After the Firefly had been run forward and the requisite landing gear of arrestor wires and crash barriers was rigged on the deck, the Seafire was sent, after all, to Point Cook air base, where it was joined by the other aircraft after their flight over the city.

Rear Admiral George Creasy, as Flag Officer (Air) Far East, and Captain Robert Dickson, DSO, entertained reporters aboard the *Theseus*, pointing out that the Australian aircraft carriers would have to be better fitted for the tropics than their own ships had been. They had suffered on the journey from Ceylon. Flight decks had become red hot and the Admiral's cabin registered a temperature of 106 degrees (41°C). 'Our sick bays were almost full in the tropics, what with prickly heat, dermatitis and boils. Now we are almost back to normal'.[1] The cold weather in Hobart had not been at all unwelcome.

Creasy also came with a message to Australia:

> Sea power is not an automatic or God-sent thing conferred upon us, we must work and sacrifice ourselves to maintain it, we are every whit as dependent on it as we ever were in the past.
>
> Sea power means exactly what it says, the power to use the sea. We must have our navies to ensure that none of our enemies can ever interfere with our use of the seas, and in turn to deny them the use of it if the time ever comes.[2]

He had been in Melbourne before, when he served aboard the cruiser *Sussex* during the visit of the Duke of Gloucester in 1934. During the war, he commanded the First Destroyer Flotilla in offshore patrol work in the English Channel. It was this unit that had rescued the Dutch Royal Family in the face of the advancing German armies. The *Theseus*' Sea Otter was declared to be his Admiral's barge. Captain Dickson had a different Australian connection. He had been a midshipman at the Anzac landings in 1915.

Aboard the *Theseus* was the 14th Carrier Airgroup — 804 Squadron of 13 Seafires and 812 Squadron of 12 Fireflies. Aboard the *Glory* was the 16th Carrier Airgroup — 806 Squadron of 13 Seafires and 837 Squadron of 12 Fireflies. 812 and 837 Squadrons had been aboard the *Vengeance* and *Glory* respectively as part of the British Pacific Fleet, though they were equipped at this time with Barracudas. The Seafire, basking in the Battle of Britain glory of the marque, was regarded aboard the ships as faster and more manoeuvrable than the RAAF Mustangs. The Firefly was intended to operate primarily in an anti-shipping and anti-submarine strike role while the Seafire provided combat air patrol and fleet air defence. Each ship had a Sea Otter amphibian, and Captain Dickson had his personal Tiger Moth. This last aircraft, when assembled for his use on approaching Melbourne, succeeded in succumbing to engine trouble.

On the Australian side, the Mustang had entered service with the RAAF in late 1944, and the first Australian-built machine flew on 30 April 1945. Compared with the Seafire, it was generally faster, more rapid in climb, and had a greater operating ceiling. It was described as 'one of the world's best pursuit planes'.[3] As a fighter it carried six .5 inch machine-guns, compared with the Seafire's two 20 mm cannon and four .303 inch machine-guns.

On a cold Sunday as many as 67 000 people were on Station Pier, 50 000 of whom got aboard the ships, which were said to have emerged 'apparently intact'.[4] Small boys were reported to be playing with the Bofors guns while their big sisters talked to the sailors. No guns went missing. Cameras, even those of the simplest type, were strictly prohibited for security reasons. Public transport was doubled for the day. Proceeds from the sales of a souvenir brochure, priced at a shilling each, went towards the Lord Mayor's Food for Britain Appeal and local charities.

Aboard the *Glory* was Captain 'Nutty' Gosling of the Royal Marines, reputedly the most beribboned man in the fleet. Alone and idle the previous night, he had rung all the Goslings in the Melbourne phone book and invited them aboard as his guests.

An Australian aboard the *Glory* was Lieutenant Brian Murray RAN, the air direction officer. He was later to finish a fine career as Governor of Victoria. The crews, some 2600 in all, were provided with lavish hospitality, dances and social occasions. One who was not given leave was a Japanese war criminal, Captain Taura Hidesharu, who was being shipped to Rabaul to commence a five-year sentence.

On the Tuesday the ships sailed for an aerial demonstration on Port Phillip. The wind was gusty and variable and as 12 aircraft flew in from Point Cook to land on the *Theseus* they had to take 25 separate landing approaches as she twisted and turned to find the wind. The *Glory* operated in the western part of the Bay while the *Theseus* was better displayed in the east. Sixteen of her aircraft flew overhead in the shape of a T, a Seafire took off with rocket assistance, and there were rocket and machine-gun attacks on a floating target, all for the benefit of the Australian service chiefs on board, including Lieutenant General Sir Sydney Rowell and Air Vice-Marshall Ellis Wackett. With a falling cloud base, the bombing display was cancelled.

Operations aboard the *Glory* were similar, with aircraft landing and, after briefing, taking off to attack the towed target. Aircraft landed at 45-second intervals. Her commander, Captain W. T. C. Couchman, had been with her since September 1946, and she was well worked-up. It was noted, however, that only five days of flying had been done since the squadron left Singapore.

On Friday 18 July, 1200 men from the ships marched through the grey overcast city to a tremendous reception. Sailors were noted buying up tinned food to take home to Britain, where things were not nearly as good as in Australia. January 1947 had seen the big freeze, the coldest weather for decades in Britain, and this, combined with considerable industrial agitation and severe rationing, made it all rather gloomy. Admiral Creasy pointed out that life back in Britain was more difficult now than at any time during the war. Sugar rationing finished in Australia in July 1947, though the supply of other items such as meat, butter and tea remained controlled.

On Sunday 20 July 10 000 people gathered to see them off, while the bands played 'Waltzing Matilda' on deck as the tugs pulled the big ships out. As the carriers sailed, they provided a flying display for large crowds at Frankston. At 1 p.m. the planes from Point Cook landed aboard, and at 2 p.m., 17 were launched again. Then, before the horrified gaze of observers, two Fireflies collided in mid air at a height of 1000 feet, soon after taking off from the *Theseus* and, locked together, crashed into the sea. The destroyer *Cockade* steamed to the scene and lowered a boat, but all four men on the aircraft were killed. One of the dead pilots had been a prisoner of war for five years. It was calm and clear on Port Phillip, and the pilots blamed the absence of wind and the position of the ships steaming into the sun for their subsequent mishaps. It is of interest to note that an Argentine vessel of the same class, the *25 of May*, was prevented from launching heavily armed aircraft for a long-range attack against British warships in the Falklands in 1982 precisely because of unusually light wind conditions. The Indian carrier *Vikrant* suffered similar difficulties in the war against Pakistan in 1971.

Out on the bay, trouble continued when a Seafire crashed while landing on the *Theseus*, killing one of the deck crew. The remainder of her aircraft were diverted to the RAAF base at Point Cook, where another Firefly was severely damaged as its undercarriage collapsed. The *Glory* was off Dromana with 20 aircraft on deck ready to take off. The Seafires went first, but on learning of the mishaps to the *Theseus* airgroup, the Fireflies were halted and run forward to the parking area. The Seafires were then brought in. One burst a tyre, but another missed the arrestor wires on landing, jumped the crash barrier and crashed into parked aircraft. One sailor was killed and another injured, and a shocked crowd on Station Pier was able to see the *Contest* arrive with the casualties. For the *Glory*, it was the first accident for 15 months, but it was not a good total for the day — six killed and one injured, and seven aircraft damaged or lost.

It was, nevertheless, intended as a showcase, and on board the ships, the Secretary of the Navy, Alfred Nankervis, and the Second Naval Member, Commodore John Armstrong, DSO, told reporters as they sailed with the squadron that the first Australian aircraft carrier, of the same type as these British ships, would commission in September 1948 or earlier, and the second would come a year later. Commodore Armstrong had served aboard the RN escort carrier *Vindex*. It was stated that the Admiralty had supplied the ships at 'considerably reduced cost',[5] and that changes were required to fit them for tropical service. This would involve air-conditioning and revised living space, as the discomfort of the British ships in the tropics had been specially noted. A Fourth Naval Member was to be added to advise on aviation, and for this purpose Captain Edmund Anstice had been lent by the Royal Navy. The RAAF would help with initial training, and recruitment would be announced soon. It was hoped to attract men with flying experience from the Royal Navy. Maybe there would be some who had formed romantic attachments while on this cruise.

The squadron cleared Port Phillip at 3 p.m. on 21 July, the *Theseus* retrieved her aircraft next day, and the exercise then proceeded. The ships involved were:

British Fleet
Aircraft Carriers   *Theseus*
                    Glory
Destroyers   *Cockade*
                    Contest

Australian Fleet
Cruiser	*Australia*
Destroyers	*Arunta*
	Bataan
	Warramunga
Frigates	*Murchison*
	Shoalhaven

Off the New South Wales coast, Exercise Billabong, the first major fleet manoeuvres in Australian waters since the war, got under way in the early hours of the morning of 23 July, when a Catalina from Rathmines located the carrier force 30 miles seaward of Kiama. Lincoln bombers attacked at dawn, and at 7.45 a.m., Beaufighter torpedo bombers with Mustang escorts roared in for an attack on the carriers, which they countered with Seafire fighters. After the initial attack with rockets and torpedoes, the British and Australian planes engaged in dogfights, and the score was agreed as one carrier disabled and three Seafires shot down. Another Seafire which had to make a forced landing at Nowra was not counted. Such was the enthusiasm, that even a RAAF Mosquito joined in. Off Botany Bay, the Beaufighters attacked again, in support of the Australian destroyers, which raced to fire torpedoes from behind a smoke screen. With the *Glory* declared disabled, the *Australia* sailed in to finish her off with a classic gunnery action in the old style. On board the *Theseus* off Jervis Bay, operations were flawless, with the Fireflies, and then the Seafires, landing at 14-second intervals.

Next day at 7.30 a.m. the carrier aircraft flew in low from the Heads, and the whole fleet sailed into Sydney harbour. Rear Admiral Creasy, while pointing out that in wartime circumstances, the carriers would not have been so recklessly exposed off a hostile coast, said, 'We learned a devil of a lot'.[6]

At this time, Field Marshal Viscount Montgomery of Alamein was touring Australia, to considerable public acclaim. He was pushing hard at the time for adequate rearmament to face a possible Soviet push in Europe. Everyone knew what he meant when he said that Britain had been totally unprepared for war in 1939, but Admiral Creasy felt that he had to add something on behalf of the navy. It was true that more warships would have been useful, he pointed out, they always are, but there was no question that the navy was unprepared, ever![7]

In Sydney the ships were open to the public, and there was a march by the cenotaph in Martin Place. To celebrate the 47th birthday of Queen-Consort Elizabeth, the ships were dressed in flags, and the *Australia* fired a salute. Sydney by this time was fairly used to visits by aircraft carriers. Ships of the British Pacific Fleet had been regular visitors up to 1946, and in May it had been host to the USS *Antietam* and *Shangri La*. On 5 August the fleet sailed for Brisbane. The *Glory*, for which this was the start of the voyage home, flew a long paying-off pennant. There were further exercises en route, in which the British submarine *Amphion* participated.

On 6 August there was more trouble when the carriers were engaged in interception practice using Fireflies. Six aircraft were flown off each ship at 10 a.m. while Mustangs from Amberley were used for the attack. The crew aboard one of the *Theseus'* aircraft found they could not retract the undercarriage and so landed again. The other five aircraft returned at 11.30, the first three landing safely. The fourth, however, bounced heavily, the second time clearing the crash barrier and landing among the parked aircraft. It damaged the wing of one, and hit another square on, so that the two aircraft went over the side. The last also bounced badly, went over the crash barrier and hit the island just below the Admiral's bridge with its wing. It came down on its nose and went over the side. The *Arunta* and *Cockade* picked up the crews from the water, but one aircraft mechanic was killed.

A total flying ban was now enforced until aircrews could get in some training on an Auckland airfield. The *Theseus* had commissioned only in February, and had carried out no flying on the way out from Britain, as she was carrying an extra load of aircraft for the *Glory*, which had been in the Pacific since the end of the war. Practice flying in the meantime had been minimal. The *Cockade* meanwhile steamed at full speed for Brisbane to get one of the injured men to hospital.

The combined fleet arrived to a quiet reception in Brisbane for the Exhibition on 8 August. It was to be in port for ten days. In those days the visit was part of a regular annual cruising schedule. Subsequently there were record crowds at the Exhibition, but large numbers also flocked to see the ships. On the Sunday, 7000 went over the *Glory*, 9000 went over the *Theseus*, and 4000 missed out. The trams were packed to the footboards. 'Young girls hurried on board to discuss the ships with attentive sailors'.[8] One opportunist was seen to be training a Bofors gun on the bookmakers' area at nearby Eagle Farm racecourse. The best thing about Brisbane for the sailors was that the hotel bars did not close at 6 p.m., as they did in the south. The only catch was that the sailors had very little money to spend, and they could frequently be found just walking up and down the streets. There was a big dance at the City Hall, at which the girls outnumbered the sailors by three to one, a march through the city of 1500 sailors, and at the

Cremorne Theatre the sailors were entertained with a performance of 'Nautical Nights', with 'Cremorne's Treasure Chest of 12 Golden Cuties and 6 Glamorous "Pin-up" Girls (all very Ship Shapely), and Ken McPherson's Band of 12 Nifty Nautical Boys!'

In Brisbane the fleet was joined by the cruiser *Hobart*, on its way south from Japan. The *Glory*, heading for England and home, via Singapore, took 7500 food parcels with her. Many more had been posted. The *Bataan* then sailed on to Japan, and the *Contest* to Hong Kong. The *Australia* went to New Guinea where she was to convey notables to Manus Island, and thereafter to Japan, where she was the last Australian cruiser to serve with the occupation forces. There, she was active in exercises with U.S. and Royal Navy cruisers and destroyers, and her gunnery and radar were noted to be in top condition. The *Theseus* and *Cockade*, for their part, went on to New Zealand.

New Zealand at this time had in commission only the cruiser *Bellona* and the corvette *Arbutus*. The *Bellona* was in the process of returning from a two-month training cruise for new recruits among the Pacific islands. She had visited Hobart in April 1947.

The British ships had an enthusiastic welcome in Wellington, and then moved on to Auckland, where they arrived on 30 August. In the early morning, off Tiritiri, preparations were made for flying off the aircraft. As the deck crew could not start the engine of the Sea Otter, she was hurriedly struck down, while the 11 Seafires and 12 Fireflies followed the Tiger Moth off the deck, dodging the rain and heading for the RNZAF base at Whenuapai. Serious training was to begin the following day on a section of the airfield marked out as a carrier's deck. Admiral Creasy was very gracious in describing the facilities of the base. 'They had seen nothing like it since they had left Britain.'[9] The recalcitrant Sea Otter was hoisted out and took off from the harbour to join the rest of her brood three days later.

Bearing in mind the limited finances of the sailors, and the grim conditions in Britain, where a single New Zealand meal would have cost a week's meat rations, the Royal Society of St George donated a food parcel for the next of kin of every man aboard the two ships. 'Many of us have very little money to buy things for our folk', said one officer, 'and a gesture like this is a very real act of kindness.'[10]

Crowds visited the aircraft carrier. Some 6000 squeezed aboard on the first Sunday, with 2000 more just on the wharf. On Saturday 13 September there was a big flying display at Whenuapai, where, despite hail and driving rain, 4000 people watched the mock attacks and aerobatics, and the flypast of 16 aircraft in the shape of a T. The Tiger Moth was meant to amuse the crowd by bursting balloons with its propeller, but the 50 mph wind was too much, and all the balloons got away.

On Monday 15 September the airgroup flew back aboard the *Theseus* in the Hauraki Gulf, and two days later she sailed for Guadalcanal. Unlike the *Glory*, which was sailing home to be laid up, the *Theseus* was to spend most of the next two years as flagship of the Third Aircraft Carrier Squadron, but not in eastern waters. The parlous state of the British economy obliged the Admiralty to announce, on 27 October, that it was withdrawing all of its aircraft carriers from the Far East Station. The result was felt in April 1949, when there was no air support available for warships which were attempting to rescue the *Amethyst* from its predicament on the Yangtse. Inevitably, policy had to be reversed. Both the *Theseus* and the *Glory* were to see action, with honour, later in Korea.

Meanwhile in Australia, a recruiting drive began to fill 448 places in Naval Aviation. There were 4200 applications in three weeks.

3

THE END OF THE BEGINNING

Things were much more efficient with the visit of United States Task Force 38, when the opportunity was taken to combine the regular squadron exercises with the participation of a major foreign force. Task Force 38 was famous as the principal United States carrier striking force in the latter days of the Pacific war, though now much reduced. The American squadron sailed into Sydney from Pearl Harbor on 30 January 1948 and the sailors were on their best behaviour, in contrast to previous occasions. It helped that they were strictly limited in the number of cigarettes they could buy on board ship for trading in the street. During their stay, there was a banquet at Ushers Hotel at which the American Rear Admiral Harold Martin had said, 'Don't keep your navy at home; send it to all parts of the world',[1] and the Minister for the Navy, William Riordan had admitted that the RAN needed training in aircraft carrier tactics. With the powerful American force, they were going to get it. Ships which took part in the exercises were:

American Fleet
Aircraft Carrier *Valley Forge*
Destroyers *Keppler*
Lloyd Thomas
William C. Lawe
William M. Wood
Oiler *Mispillion*

Australian Fleet
Cruiser *Australia*
Destroyers *Bataan*
Quiberon
Quickmatch
Frigates *Culgoa*
Murchison
Shoalhaven

British Vessel
Submarine *Aeneas*

The *Valley Forge*, a very much larger and faster vessel than the Australian carriers were to be, was of a type that carried between 87 and 104 aircraft during the war. She carried Airgroup 11, comprising Bearcat fighters, Avenger torpedo attack aircraft and Helldiver bombers. The Bearcat, which had first entered squadron service in May 1945, was a high-performance interceptor aircraft, regarded as the navy's best, though it was soon to be overshadowed by new jet aircraft. It was a 'hot rod', noted for its phenomenal rate of climb compared with, for instance, the Mustang. The *Valley Forge*, which had commissioned in February 1946, had undergone a four-month period of intensive training since being homeported to San Diego and her crew were at the top of their form. Her commanding officer was Captain John Harris. Admiral Martin was a veteran of the Pacific war, with service including command of a force of seven escort carriers in the assault on Okinawa.

One pleasant formality took place aboard HMAS *Australia* when Rear Admiral Martin presented Rear Admiral Farncomb with the U.S. Navy Cross for his actions as commodore commanding the Australian Squadron in the invasion of the Philippines in 1945.

The combined force sailed from Sydney on 4 February, and the carrier worked up to 29 knots to fly off aircraft. In the first stage, RAAF Mustangs from Williamtown protected the Australian ships from American attack, while Beaufighters from Richmond and Lincolns from Amberley took the offensive. Some hits were claimed on the carrier and some aircraft were 'shot down'. At night the fleet was shadowed by Catalinas from Rathmines, though one was claimed as a casualty, and at first light on 5th, the U.S. Bearcats were divided, some for attack and some for defence, while the Beaufighters came in again for successful torpedo attacks. The *Aeneas* exercised separately with the *Murchison*, which was trying out her hedgehog anti-submarine weapon for the first time. The submarine later made a successful attack on the *Australia*, even though she was screened by the three frigates. The American force proceeded then on its way to Hong Kong, while most of the Australians sailed south for Westernport. The *Australia* and *Bataan* were to continue to Hobart for the Regatta and then to Dunedin as the first part of a visit to ports in New Zealand. On leaving Auckland on 30 March 1948, the *Australia* and *Bataan* exercised with the lone New Zealand warship, *Bellona*.

The cruise of the *Valley Forge* subsequently included operations in the Persian Gulf, for the purpose of impressing Crown Prince Saud. Despite bad weather, which prevented the proposed flypast with the aircraft forming the letters SAUD, there was some flying. Not the least impressive aspect of the visit, from a practical point of view, was a showing of the 1945 documentary film shot aboard USS *Yorktown*, *The Fighting Lady*.

The U.S. command afloat in the western Pacific was to be designated the 'Seventh Task Fleet' in August 1949. The *Valley Forge* was to be the first U.S. aircraft carrier in action in Korea, on 3 July 1950.

1948 was the real start of the Cold War, a term that had been coined in the U.S. early the previous year. The Communist coup in Prague in February and the start of the Berlin blockade in July seemed to be setting a pattern for the future, and from an Australian perspective, Communist activity in Malaya, Vietnam and China could not be ignored. Once again, Australian eyes turned towards an Asian threat and commentators noted the cutback of British forces. They were far from sure that Britain was right in 'gambling on a period of peace'.[2] Defence was going to be an important priority.

One curiosity of the time was the demise of a previous generation of Australian naval aircraft. The Antarctic expeditions of 1947-48 used the old polar vessel *Wyatt Earp* and *LST 3501*, the former carrying a Sikorsky Kingfisher seaplane and the latter a Walrus amphibian. Heard Island was, on the occasion of this visit, transferred from British to Australian sovereignty on 26 December 1947. There was much bad weather and the Walrus, the last in Australian service, was blown to pieces in a violent gale in mid-December. On Heard Island, some of the crew painted a suitable rock orange, as the egg of the 'Pusser's Duck', but as it proved infertile and the breed was declared extinct. Meteorological stations were established at Heard and Macquarie Islands as was a fuel dump at the French island of Kerguelen. In late 1948, *LST 3501* was named *Labuan*. She was to remain the major element of the navy's support of the sub-Antarctic bases until 1951.

The training of Australian pilots for naval aviation began at Point Cook in December 1947. Starting on Tiger Moths and Wirraways they graduated to RN Seafires or Fireflies. In 1948, Tiger Moth, Wirraway and Spitfire aircraft were transferred from the RAAF to the RAN for training purposes. To exercise command over naval aviation, Commodore Guy Willoughby, RN, was appointed as the first 'Commodore, Air'. Captain Roy Dowling, who was to commission the first ship, left in February along with senior specialists for training in the U.K. Men who were transferring from general service to the air branch did courses at Sale. Alan Porter was one who emerged as 'naval air arm photographer'. Some of the 1947 course, including Noel Knappstein, went to England to work up aboard the fleet trials and training ship *Illustrious*.

After the loss of the cruiser *Sydney* in 1941, a Replacement Fund had been established by public subscription, and its £427 000 transferred to help pay for one of the new aircraft carriers. On 27 April 1948 it was announced that the names of the new Australian ships would be *Sydney* and *Melbourne*, and the naval air station at Nowra *Albatross*, thus perpetuating appropriate former names from the fleet.

On 23 June the transport *Kanimbla* sailed from Melbourne with 500 men as crew for the *Sydney*. She also carried a cargo of rice as well as food gifts for the people of the naval towns of Devonshire. The *Sydney* was due to commission in October or November, and the *Melbourne* was due to be completed in 1952, to complete Australia's carrier task force. The aircraft carriers of the class were designed to have a full war complement of 1343 men, but in Korea, the *Sydney* was to carry 1427. In the event, the British shipyards were the cause of serious delays to the program. This was partly due to a re-allocation of priorities towards the rebuilding of the merchant fleet so badly depleted by the war and partly to the bloody-mindedness of the British workers.

British shipyards were utterly crippled with steel shortages from 1947 to after 1951. As a result, shipbuilding fell well short of government projections and the yards often found themselves operating at half capacity. Order books were full as 'owners were gasping for yards',[3] but delays continued, and costs rose. During the time that the *Melbourne* sat in the yard at Barrow, she would have seen a new generation of ocean liners pass by which were to become household names in our part of the world: *Orcades, Himalaya, Chusan, Oronsay, Orsova*; all were laid down and completed while the carrier was brought on at a snail's pace.

The *Sydney* herself was delayed, and in the interim, her crew were housed at Devonport in the *Glory*, which was laid up pending refit. The Australians found her in a very dirty state, and were subsequently very proud of the way they kept the *Sydney* clean. Many were bemused by the work being done removing the experimental flexible deck from another British carrier, the *Warrior*. This had been for possible use with jet aircraft and seemed the strangest thing to the average sailor.

For her return voyage to Australia, the cargo of the *Kanimbla* included two turrets for the new destroyer *Tobruk*.

It was a great day, 28 August 1948, when the 20th Carrier Air Group of Naval Aviation, RAN, comprising 805 Squadron of 12 Sea Furies and 816 Squadron of 12 Fireflies, were commissioned at RNAS Eglinton, in Northern Ireland. Like ships, the squadrons brought their British battle honours with them when they entered the Australian service. For instance, 805 had Crete 1941 and Libya 1941-42. After the pilots completed their basic training in Australia, they had graduated to the U.K. operational flying school, RNAS Lossiemouth, in Scotland and then to RNAS Eglinton in Ireland.

Albatross, the shore station, was commissioned on 31 August 1948 and the *Sydney* was handed over at Devonport for trials on 16 December.* The first aircraft carrier to be built and completed in a Royal Dockyard, she was given her name by Mrs J. A. Beasley, wife of the High Commissioner. Captain Dowling had previously commanded the *Hobart*, from December 1944 to April 1946, and prior to that had been in an RN carrier during the invasion of southern France.

The title of His Majesty's Australian Fleet was granted to the Australian navy on 1 January 1949, in recognition of its new strength and aspirations, with the fitting out of the aircraft carriers.

The first aircraft to land on the *Sydney*'s deck was an American helicopter, on Christmas Day 1948, but two aircraft were soon borrowed for training. On her full power trials on 6 January she recorded 24.61 knots on the measured mile. It was anticipated that, six months out of dock in the tropics, and in deep load condition, she would lose three knots.

After her trials, the *Sydney* was accepted on 5 February 1949 and proceeded to embark aircraft: first a number of British machines for trial purposes, and then her own squadrons, from the Royal Naval air station at Arbroath in Scotland, where there had been six weeks of intensive flight training.

Initial training was followed off the Cornish coast by tactical exercises in search and strike with Coastal Command. There was only one serious accident in 686 deck landings. Off the Scottish coast there were gales and snowstorms — a presage of conditions that would be experienced later in a real war. Flying from a deck covered in ice was very tricky. One strike, by Fireflies in the Western Approaches, was on the liner *Queen Mary*. The destroyer detailed as plane guard through all of this was HMS *Contest*. *Sydney* was the beneficiary of a change in Royal Naval flying doctrine resulting from unacceptable accident levels during the late 1940s. It was just not good enough to accept the current British average of one accident in every 50 deck landings, and a system more akin to that in use in the U.S. navy was adopted. This included different approach techniques and bat signals, and by 1951 the accident rate in the Royal Navy had been halved. In *Sydney*'s case, as with all new carrier groups, there would be accidents until everyone was used to the aircraft and the ship in most conditions.

At Milford Haven, in Wales, the wind was blowing so strongly that a Firefly was able to take off while the ship was still at anchor. Just one story

---

* A small incident had threatened to mar the proceedings when, a few days previously, the ship's engines were damaged as the dockyard engineers attempted to start them. They found seven steel bolts in the sump of the gearbox connecting the turbine to the starboard propeller. Sabotage was feared.

should be told of one of the more spectacular accidents. A Firefly was coming in but the batsman gave it the waveoff. The pilot, Danny Buchanan, increased revs and went into boost but he neglected to attend to his propeller pitch and to adjust his air brakes. The net result was that he 'did a bellyflop' right among the twelve aircraft parked forward and wrote off seven of them. In one, the observer was still sitting in his seat. By very great fortune, there was nobody killed and no fire, but, as Alan Porter remembers, 'just a bloody big silence for about 20 seconds, then someone said 'Sh★★!" (Traditional sailor's expression, short for 'Shiver my timbers!')

The Australian sailors were grateful to be able to buy ship's stores to take as gifts when they visited English families as rationing was still very severe. In Ireland, on the other hand, food was plentiful though expensive, but naval condoms, issued free on the ship as 'wet weather gear', brought a good price on the black market. There was one more thing. A syndicate of the aircrew had a big win on the football pools and converted the currency into 20 motor cars to bring home to Australia. She was, as Captain Dowling said, 'a happy ship'.[4]

On 12 April 1949 the *Sydney* sailed from Devonport for home. She carried a total of 52 aircraft, of which 28 were spares, and her normal complement was inflated to 1635 for the journey by 600 extra ratings carried as passengers. Also aboard was Harry Fitzpatrick, Senior Inspecting Officer of RAN Stores, who had been in England studying the requirements of aircraft carriers. She made her first Australian landfall at Fremantle exactly a month later, where she was greeted by the grand old lady of the fleet, the cruiser *Australia*, along with the destroyer *Warramunga*. Now 21 years old, the cruiser had recently indulged in the first practice shoot of her main armament in over a year, and her gunnery had been rated very poor.

It was a time of great tension as the communist forces swept towards victory in China. On 19 April the British frigate *Amethyst* was badly damaged and trapped, hostage to communist guns, on the Yangtse. It was only good fortune, and Australian operational policy, that the ship involved had not been the Australian frigate *Shoalhaven*. Indeed, early in May the British authorities in Hong Kong were believed to have requested that the *Sydney* be sent there immediately on her arrival, to protect the colony against possible communist attack, prior to the arrival of RN carriers from the Mediterranean. The *Ocean* arrived in the colony with a deck load of Spitfires in August and the *Triumph* was sent to restore the proper level of naval force in the Far East, along with the support carrier *Unicorn*, in September.

Australia avoided the entanglement of its forces in the indefensible Hong Kong. Late in May, Britain wanted the aircraft carrier from Australia, as well as an infantry battalion and a fighter squadron. It was a big ask. What Britain was actually given was some shipping space for its own troops and nothing more.

On Wednesday 18 May a crowd of 5000 people gathered to see the new aircraft carrier berth at Station Pier, Melbourne. The ship's band played 'Ship Ahoy', and a squadron of 10 Mustangs flew over in salute, accompanied by an RAAF helicopter and a Vampire jet. Captain Dowling's family joined him as the ship made her way up the Bay in the early morning mist. 'On mirror-like seas, in the wake of the fussing tugs, her huge grey bulk loomed out of the morning fog.'[5]

It was school holiday time, so many of those waiting were children, and there was an air of expectancy until the first 'Coo-ee' rang out, followed by shouts and cheers. Many sweethearts and wives were present who were naturally very pleased to have their men back after a long time away. Captain Dowling, noting that he had joined the navy 34 years previously as a cadet midshipman, stated: 'This is a very proud and happy day. Many of us have been away for a year, and now we have the honour of bringing back Australia's first aircraft carrier.'[6]

With headlines in the daily papers chronicling the fall of Shanghai to the Communists, *The Age* editorialised:

> H.M.A.S. Sydney brings to the R.A.N. much more than the first of a series of modern ships of war with modern fighting aircraft. By enabling Australia to play a better role in regional defence in the Pacific, aircraft carriers based on Australia will strengthen the role of the British Empire in global defence...
>
> In welcoming the ship, Australia welcomes her returning native sons, and also her adopted sons, not only as participants in Commonwealth and Empire defence, but as citizens of Australia.[7]

Much was made of the comfort of the new ship. This was enhanced by Commander Beecher's Welfare Committee, which had, for instance, coped with the ratings' complaints about too much stew by increasing roast dinners.* Unlike the British carriers, the *Sydney* had cafeteria messing. A big crowd admired the ship and aircraft on the Sunday, and 3000 were there to see her off on 23 May.

She disembarked her aircraft to lighters at Jervis Bay on 25 May, and they were then towed up the road to the naval air station at Nowra. Captain Robert Poole, RN, commanding the base, had 500 officers and ratings, as well as 200 men from the Department of Works, putting on the finishing

---

* The Australian experience is in contrast to that of Canada, where three ships without welfare committees suffered mutinies in 1949.

touches. The shabby wartime huts were repaired, and new roads and buildings were added. There was a galley capable of serving 1000 men, and an 800-seat cinema. Radar was to be installed on a neighbouring hill. Everyone was keen that refresher training for the pilots should begin as soon as possible.

First Naval Member, Rear Admiral John Collins was becoming anxious about manpower, which currently stood at 10 450, including 4040 seagoing. He hoped that by 1952 it would have reached the target figure of 14 753, including 6756 seagoing. A thousand former RN personnel were sought but recruiting on the whole had been disappointing, and it was considered that further inducements were needed. Collins felt also that he had to defend the need for carriers:

> A fleet that goes to sea without its aircraft to-day is just as obsolete as a fleet under sail.
>
> Carrier-borne aircraft are more effective than those based ashore because they are at the immediate command of the admiral.
>
> An admiral has no need to break wireless silence to call up planes which he frequently finds have already been allotted another task.
>
> Carriers are the most important fighting ships and striking force of a fleet and are not outmoded despite the development of guided weapons and atomic warheads.
>
> Carriers give a fleet tremendously increased striking power and widely increased mobility — factors essential to success in present-day warfare.
>
> A balanced fleet to-day must have carriers as its striking force, capital ships or cruisers to protect the carriers from surprise attack when aircraft cannot be launched, and a destroyer screen.[8]

His was not the only voice. Navy Minister Riordan had no doubts when he spoke in the 1948 Budget debate. After listing the disasters suffered during the war by navies unable to gather together a balanced force, he said:

> The modern development of the Fleet Air Arm is as revolutionary as was the establishment of the Royal Australian Navy in 1911, and its introduction means that the Royal Australian Navy is being moulded into a fighting force of the most modern design and that its efficiency is being increased enormously.[9]

The *Sydney*, third bearer of a proud name, entered her home port on Saturday 28 May 1949 where 2000 people gathered at Garden Island to greet her. As she rounded Bradleys Head, due honours were given to the mast of her illustrious predecessor, the first *Sydney*, standing on top of the old forts. The ratings all had addresses for 'food for Britain', and there was a march of 377 men through the city streets on Friday 3 June. Sydney, unfortunately, was a bit blasé about the fleet, and the people who saw them pass were described as 'wildly unenthusiastic'.[10] As one of the sailors was heard to remark, you would have thought that the name of the ship was *Melbourne*, for the contrast in the reception given to it in the two cities.

It was a well-timed arrival. 1949 was the year of the greatest communist successes in China; the year, too, of the escape of the *Amethyst*. Where, some asked, would it end? It was good that Australia at last had its aircraft carrier.

4

HIS MAJESTY'S AUSTRALIAN FLEET

By the end of 1948, most of the left-over chores of war, minesweeping, disposing of unwanted ammunition, sale of surplus vessels, etc. were complete, and the navy could settle back into a peacetime routine. There had been one major accident, in which a minesweeper had been sunk. On 13 September 1947 the Twentieth Minesweeping Flotilla had been working near Cockburn Reef, off Cape Grenville, north of Cairns. They were working in an echelon line, with each vessel protected by the sweeps of the one ahead as they entered the minefield. *Swan* was in the lead when she lost a sweep with a paravane and a long cable. *Warrnambool*, next astern, passed astern of her, but in doing so, hit a mine and sank. Four were killed and 29 injured.[1] The work on the Barrier Reef passages was finished in 1947 and the flotilla was then reconstituted to comprise *Swan, Kangaroo*, two GPVs and two motor launches, for work sweeping magnetic mines in New Guinea. After that, the flotilla came home and paid off in August 1948.

In the course of its year's cruising, the squadron would, as likely as not, be in Hobart in February for the Regatta, and between training exercises in the Jervis Bay area and refitting in Sydney, there would be the Spring Cruise to Queensland, often visiting Brisbane in Exhibition Week in August, and arriving in Melbourne later, in November, for the Cup.

The fleet was kept in condition with exercises, such as those of March 1949, when the squadron was in Tasmanian waters. The *Australia* and *Bataan* then exercised with the air force in Port Phillip and Westernport Bays. There were dive-bombing attacks, the detection of attacks by Beaufighters by radar, and the vectoring of Mustangs into the defence of the ships. For the annual gunnery school firings, the *Bataan* engaged a target towed by the *Gladstone*, as well as sleeve targets towed by aircraft.

The rather leisurely and predictable seasonal movements of the Australian warships might seem a little quaint to modern observers, especially as it was evident that any war would be fought, as it had been not so many years before, in the waters to the north of Australia. Nevertheless, it had several important aspects. First, it was a time when the fleet was being built up, and recruiting was important. Its visibility around the big cities of the nation, most of which were in the south, was therefore quite important. The second aspect was that it was never too far away from its only major base, at Sydney. Third, by simply cruising together, it proved that it was a fleet in its own right and not just, as it had been during the war, a series of detached units supplementing someone else's force. Thus there was an incalculable gain, badly needed, in morale and confidence, that this was, indeed, His Majesty's Australian Fleet.

What was needed next was to exercise the fleet in conditions approximating real warfare. In 1949 Admiral Vian had commented on similar unrealities in the Royal Navy's exercises:

> Pre-war cruises were too stereotyped; economy undoubtedly prevented the Navy and Admiralty departments from becoming, and remaining, trial-minded. Exercises at sea and tests and other trials were not pushed beyond the point of minor damage. They generally failed to reveal unsatisfactory living and working conditions for war service. Thorough closing down of our ships for weeks on end would inevitably have revealed how much (and why) operational efficiency would be lost under heavy weather, arctic and tropical conditions.[2]

How Australia's forces would develop would remain to be seen. This was the active fleet in 1949:

- 1 aircraft carrier
- 1 cruiser with 8 inch guns
- 5 destroyers
- 3 frigates
- 2 landing ships (LST)
- 2 survey ships
- 1 ocean-going tug

Apart from squadron cruises and exercises, other ships of the fleet would be sailing for more widespread commitments, including occupation duties in Japan, and the northern patrol in New Guinea. The occupation force was scaled down from two ships to one in July 1948, and as time progressed, the northern patrol became rather patchy. In this latter case, it was admitted that the frigates were too large to be really useful in what was essentially a fishery patrol. The skipper of an errant fishing boat could often sight the top hamper of a frigate in good time to make himself scarce before the ship itself arrived.

On 1 October 1948 the RAN assumed responsibility for the British Commonwealth naval representation in Japan. Ships would find the shore station, HMAS *Commonwealth*, at Kure. Duties in Japan included the Kyushu patrol, intercepting refugees from Korea. In January 1949 the *Warramunga* rescued survivors of the Chinese ships *Tai Ping* and *Kien Yuan*, which had collided off the mouth of the Yangtse and sunk, plunging over 1000 people into the icy water. Later that year the *Culgoa* was the first ship to pass through the Straits of Shimonoseki following the clearing of the minefields.

A 25-year project had been set in hand in June 1946 to survey the whole Australian coast for the first time since Flinders. The *Warrego*, with her tender *Jabiru*, and the *Barcoo* were engaged in survey work, mainly in the approaches to major ports. In 1948 they had surveyed southern waters in preparation for the Royal Tour, and it was during this process, in April, that the *Barcoo* was blown ashore on the beach at Glenelg. Fortunately, she was got off without damage. The *Lachlan* also surveyed southern waters in the summer, and she and the *Barcoo* moved to north-west Western Australia for the winter. It was a big embarrassment that an aircraft carrier would be unable to manoeuvre in the waters inside the Great Barrier Reef because of inadequate charts.

The ship that never came was the battleship *Vanguard*, last and mightiest of her line in the British navy. She was due to bring the King to Australia, and to arrive in Sydney on 4 April 1949, but his ill-health caused the complete cancellation of the tour. The cancellation denied the *Australia* the honour of acting as Royal Yacht in the Great Barrier Reef. The Royal Tour was rescheduled for 1952, using the liner *Gothic*.

On 8 February 1949 the Australian advance party left for Manus Island to commence work on the new base. The LST *Tarakan* sailed from Sydney with equipment and, with the assistance of the frigate on the station, *Culgoa*, she would transfer gear from Dreger Harbour. The *Reserve* would tow the floating dock. The Minister for Defence pointed out that the base would serve as a screen for New Guinea and adjacent islands, as well as the mainland, and that, in spite of its isolation, it was comparatively easy to defend against raids because the main harbour was completely enclosed by sheltering islands and reefs. By January 1950 it was to be sufficiently advanced to be commissioned as HMAS *Seeadler*, though the name *Tarangau* was transferred after a few months, with the closure of the old facility.

Manus had been a big base when big bases were needed for the huge forces the Americans had wielded in the Pacific war. As headquarters of the commander of the U.S. Third Fleet in September 1944, there had been accommodation for 37 000 people, airfields and all other facilities, and anchorages for over 260 ships. The British Pacific Fleet had used it as a staging point on the way to and from Sydney. All that was over now. By September 1946 the Americans had decided they had no further need of it. If they wanted an island, they had plenty of their own, including the facilities at Pago Pago in Samoa.

However, Manus Island was regarded as a key to control of the sea routes in our part of the world, as necessary to Australia as Gibraltar was to Britain. Admiral Sir Louis Hamilton called it the Scapa Flow of the Pacific. Richard Casey said: 'It was not only a naval base, but a military and air base on a grand scale. Gibraltar, in comparison, was a flea-bite.'[3]

When seen by journalists in 1949, it was evident that the 'No.1 Bastion' would require a very substantial amount of work. Although the equivalent of £120 million had been spent on the base during the war, much had succumbed to tropical rot. Everything over a huge area was either falling down or being overtaken by the jungle. The main wharf needed replacing, and there were immense piles of former U.S. transport vehicles which had been bought by the Chinese but not removed. What they could not take, however, they had effectively destroyed. Journalists saw the *Malaita* unloading stores, but what most impressed them was the view of vast numbers of quonset huts, no less than 1600 of the larger type, that made up the former U.S. facility. More than 300 Works and Housing employees, including 200 refugees from the Baltic states, were in residence, working to transform the base, and it was expected that 90 naval personnel would soon be housed, and that the number would grow as things developed.

The LSTs were occupied in taking motor transport, earth-moving equipment and stores of all kinds north to Manus. They returned loaded with salvaged U.S. items for use in Australian housing schemes.

Assisting in all the work were a selection of Japanese war criminals with sentences from five years to life. They were held in a former U.S. Marine detention compound, and they worked 10 hours a day. War crimes trials continued in the islands until mid-1951. Those at Manus Island itself, which involved some 120 Japanese, commenced in June 1950 and extended to April 1951.

The Australian defences, nevertheless, were the subject of some critical scrutiny from the U.S. In particular, they said, 'Australia will be virtually defenceless for four to five years, and the Manus base cannot be effectively manned'.[4]

Rear Admiral Harold Farncomb echoed the sentiment from his perspective as commander of the Australian fleet. He pointed out that you could build ships, tanks and aircraft, but they were useless unless you could man them. Australia was enjoying a golden age in his opinion, and the three services had to compete for manpower with industry. The 40-hour week was introduced at the start of 1948.

The matter of the Manus Island base was a complex one for the navy, and it was a problem which was never actually resolved. First, it was an exposed forward base, as Subic Bay had been for the Americans in the recent war, and therefore its viability in war conditions relied on the availability in the vicinity of a strong force of warships. Indeed, unless major units were homeported at Manus, it would remain a liability, always potentially open to enemy attack. Australia was to have its strong fleet, with its two aircraft carriers, its cruisers and destroyers, yet in reality it never could and never intended to base the carriers on the island. Neither Britain nor the U.S.A. ever considered Manus seriously after the war, as they were able to move back to Singapore and Subic Bay, where former extensive base facilities could be repaired and recommissioned cheaply. Indeed, so despondent had some British planners become in the climate of postwar economic stringency, that they could envisage the Singapore base turned over to the Australian commander, who would supply, from the RAN, its principal forces. If this were ever to become reality, a major base at Manus would be redundant.

From another point of view, Manus was not adequately defended to supply a fleet with a secure base. The Australian army had no intention to mount any coastal defences on the island, even though the Australian coastal gun batteries elsewhere were to be maintained for another decade.

Manus was to become part of a forgotten dream.

The *Tarakan* was to finish her association with Manus rather abruptly. On 25 January 1950, while she was alongside at Garden Island prior to sailing north, she suffered an internal explosion which killed seven sailors and a dockyard worker, and injured a further 13. She never went to sea again.

For the navy, there were other duties to attend to further south. The *Labuan* had to support the annual expedition to Macquarie and Heard Islands. People who sailed in her remember how her long deck used to flex as she met the big swells. *HDML 1328* went recruiting in the New South Wales outports, and her tiny crew were rather overwhelmed by the ceremonial welcomes accorded them at Lismore and Grafton.

In July 1949 the crew of the *Warramunga* helped save Melbourne from the worst of the coal strike, by replacing the wharf labourers. In this case, the *Haligonian Duke*, with 6700 tons of coal, had been black-banned for 157 days. The coal strike was a serious crisis, and there was a real chance of a collapse of the Melbourne gas supply. The navy would have manned the

Dalby, which was to carry coal cut by the army from Newcastle to Hobart, but work bans were lifted just in time. In mid-August men from the shore base HMAS *Lonsdale* took over the Melbourne tug fleet so the *Eagle* could bring the *Saint Gregory* alongside the gasworks with 6000 tons of Indian coal. In a celebrated incident of another kind, the *Australia* herself made a dash to attend a medical emergency at Heard Island in early August 1950. The passage from Melbourne took 11 days, with a return of six days back to Fremantle. The cruiser suffered from gales up to force 11 and freezing conditions, but came through with flying colours.

In the case of the new Battle class destroyers, there had been teething problems with the gunnery systems, which were close to the leading edge of British technology. The 4.5 inch guns were radar-controlled, through a primitive computer called the Flyplane Mark 3 predictor. It gave full anti-aircraft and surface control. Likewise, the STAAG mounting for the 40 mm guns, which provided automatic radar control, was found to be very sensitive.* Over time, their difficulties were fully overcome. One trap for the unwary was the tendency of the radar controlling the guns to 'walk up the wire' from the towed target towards the towing aircraft. It would not be prudent to shoot down the target tug.

The workup of the 20th Carrier Air Group commenced late in July, but a story got about that on 1 August that exercises had to be abandoned as the ship had not enough speed to launch her aircraft. It was an accusation that hurt. In war service it was found that, fully armed, a Firefly needed 21 knots of wind across the ship's bow to take off, and a Sea Fury, 28 knots. For normal training, therefore, the *Sydney*'s 24 knots should have been adequate. Despite this, rebuttals were eventually required from the highest levels. Mr Riordan said that flying was avoided when there was no wind. 'Until the 20th Carrier Air Group of the R.A.N. has returned to a fully operational state of training it is prudent to operate the aircraft from the deck with as big a margin of safety as possible.'[5]

For some good publicity, the whole airgroup was shown off in a flight across the city when the *Sydney* and *Warramunga* returned to Sydney on 12 August. Later, off Lord Howe Island, a Firefly developed a petrol blockage in one tank, which destabilised its weights sufficiently to make it dangerous to land back on the carrier. Lieutenant Peter Lowndes brought the aircraft in safely to the nearest airfield, at Kempsey. Then, at the end of August, the ship was opened to public view at Garden Island for the first time (no cameras allowed).

---

* STAAG stands for Stabilised Tachymetric Anti Aircraft Gun. Tachymetric referred to the ability to predict target position based on angular speed.

Allan Porter, as photographer on board the ship, sometimes had to take movie film of aircraft landing so that pilots could see whatever errors they might have made and correct them in future. On one occasion he was standing in a sponson with the cine camera resting on the deck so that it would be firm as he tracked it with the movement of the aircraft. Suddenly he realised that he was no longer moving the camera, which only meant one thing — the aircraft was coming straight at him. Very hurriedly he dropped the camera and himself to the bottom of the sponson, as the aircraft came to a very untidy rest on the edge of the ship. The air was just a little blue.

In mid-September the *Sydney* and *Warramunga* were in New Guinea waters, where they were joined by the *Australia*, and also by the *Bataan*, which was returning from occupation duties in Japan. The carrier had 97 officers and 1000 ratings, compared with a destroyer's 14 officers and 250 ratings. Operations on board the carrier were regarded as very efficient, and there were never less than 75 per cent of her 25 aircraft available for flying. Operations had included the attack on a wreck on Bougainville Reef with live rockets, and mock attacks by aircraft on the ships of the fleet. The ship was well thought of by her crew as 'clean', with no accidents in 800 deck landings by the time she reached Brisbane. Only once had Lieutenant D. L. Crofts, as 'bat officer', had to jump into the safety net to avoid the wing of an aircraft landing out of alignment. Of the 35 men of her airgroup, seven were Australians, including two former RAAF, four were New Zealanders, and the remainder RN — 19 pilots and observers, and five aircrewmen. Later, during her time in Korea, it was noted that 75 per cent of her airmen had flying experience in World War II. Admiral Farncomb had been the only Australian officer to command an aircraft carrier during the war, the *Attacker*, whose Seafires had helped support landings in southern France.

The *Sydney* made her first visit to Brisbane, accompanied by the *Warramunga*, from 26 to 29 September. The *Bataan*, heading for home, had called briefly to refuel on 25th. On the wharf at Hamilton to meet the ships was Admiral Farncomb's proud mother, aged 86, who had come down from Toowoomba. This was to be his last tour afloat, before going to an important position in the United States. In talking about the *Sydney*, he spoke of the future operation of jet aircraft with the fleet, but pointed out that their fuel consumption at low altitudes was high, and resulted in low endurance. In his opinion, it was better at this stage to have aircraft that were slightly slower, but of better performance. To those who accused the ship of being obsolete because she was too slow, Farncomb said they were 'uninformed':'The Sydney is a light fleet carrier, and therefore not as fast as a fleet carrier. The speed, however, is quite adequate for operating aircraft, particularly when engaged in operations.'[6]

The visit of the ship was a good time to blow the recruiting bugle in big newspaper advertisements. Small boys of the time could live out their naval fantasies with a cheap cut-out cardboard aircraft carrier, complete with aircraft.

In the matter of jet aircraft, although the Royal Navy had been conducting trials since 1945, its first operational squadron was not at sea until 1951, and none were used in Korea.

At the end of 1949 the Australian Sea Furies were given minor modifications, as recommended by the RN. The repair and overhaul of all of the naval aircraft was in the hands of Fairey Aviation Company Australia Ltd in Sydney.

In the election campaign that was to sweep Labor from power, much was made of the explosion of the first Russian atomic bomb, and accusations were levelled at the Defence Minister that he was not prepared to defend Australia. Some sections of the press were loud in accusations that Labor had invested in an aircraft carrier of obsolete design.

In outlining the naval aspects of the last Chifley Government budget, Mr Riordan attacked the 'fireside strategists'. He said that the *Sydney* had been bought for £3 million and, although a compromise between force and economy, was the most modern of her class in commission:

> Light fleet carriers are now included in all navies of any size, including the Royal Navy, and there are numerous vessels of the *Sydney*'s class in commission. Criticism has been levelled at the alleged lack of speed of the vessel, but its speed is sufficient to launch aircraft under almost any weather conditions. In addition, the vessel is fitted with a catapult to launch aircraft when conditions are such that they cannot be launched in the normal way.[7]

He concluded that *Sydney* was 'capable of fulfilling every task that it was built to undertake'. The naval development program of his government comprised:

1. Building up a destroyer fleet
2. Purchase of carriers
3. Establishment of an anti-submarine school
4. Borrowing of submarines from Britain
5. Construction of an advanced naval base at Manus Island[8]

The matter of submarines was seen as of growing importance, as new boats were faster underwater and had greater submerged endurance than those seen during the war. Mr Riordan admitted that Australia's big coastal cities could be vulnerable in the future to attack by atomic mines from Soviet submarines. The first priority therefore was anti-submarine training for the ships of the fleet, and he gave as his opinion that Australian submarines of the future would themselves be armed with atomic weapons.

Official policy was that new weapons would be introduced in an evolutionary way. 'There will be no rapid development which will render vessels such as carriers, cruisers and destroyers obsolete within the near future'.[9] Australia, like all of the dominions, was being gently edged by British Admiralty policy into the creation of an anti-submarine force. Indeed, anti-submarine warfare was seen as the only realistic role into which the dominion navies should be placed. This was for the big show; the coming world war against the forces of communism, as seen from the centres of power in another hemisphere. If local requirements seemed to demand another strategy, then Australia would have to rely on itself, and not on Britain. On the other hand, if there were to be an effective anti-submarine force, there would have to be submarines available on which it could hone its skills.

1949 ended with the arrival of the first permanent submarine presence in Sydney since the Great Depression, the Royal Navy's *Telemachus* and *Thorough*. The Fourth Submarine Squadron was to be based in Sydney at HMAS *Penguin* at Australian cost. A permanent flotilla presence was far preferable to the previous practice of occasional visits for anti-submarine training, even if the boats had to return to Singapore for refit. Each submarine had a crew of 60, and there was a spare crew provided, as well as 15 base maintenance staff. As the submarine had been identified as the major threat in a future naval war, their presence was necessary in training the Fireflies from the *Sydney*, as well as the frigates. The third submarine of the squadron, *Tactician*, arrived in July 1950. It had originally been proposed that New Zealand should share in the cost of the squadron, and that it should assist in the training for the New Zealand frigates.

On 7 June 1950 the *Sydney* sailed for Britain to embark the remainder of the navy's aircraft at Portsmouth. They comprised 59 aircraft and 250 men of the 21st Carrier Air Group, 808 and 817 Squadrons. The two squadrons had been commissioned at St Merryn in Cornwall on a significant date, 25 April 1950. The flyers would have undoubtedly have remembered that this was exactly 35 years since the landings at Gallipoli. The men of the squadrons had been training at Royal Naval Air Stations, and on joining the ship at the end of August, they worked up off the Scottish and Irish coasts in not unexpectedly foul weather. One aircraft was lost early when a Firefly bounced too high, went right over the bow, taking another aircraft with it, and was run down by the ship. Pat Hanna, the pilot, with great presence of mind, closed his canopy as the aircraft hit the water and then listened as the ship passed over him. When he heard the screws pass, he opened the canopy and bobbed to the surface astern. He suffered burst eardrums, but it could have been worse. The attendant destroyer *Wilton* then deftly picked him up. He did not fly after that but saw service as gunnery officer.

The Fireflies of 817 Squadron were the Mark 6 variant, intended for anti-submarine work. Accompanied by the destroyer *Rapid*, the airgroup was worked up in anti-submarine exercises with a Royal Navy training squadron based at Londonderry.

Captain Harries' experience handling the ship in a gale near Cape Wrath, where the seas had buckled part of the bow, was to stand him in good stead later, in a typhoon off Japan. The light fleet carriers were noted to pitch heavily in a head sea, due to their fine lines, but they did not roll heavily nor did they heel unduly under full helm. Some who had seen the *Glory* in a typhoon in 1946 remembered how the seas had come over the bow and had rolled along the deck, forcing water down the edges of the lifts with the strength of a firehose.

It was always a source of great pride to the men of the *Sydney* to be closely associated with the Royal Navy, and accepted as equals. The *Sydney*, along with the *Implacable* and *Vengeance* and other RN warships, was crowded with visitors at the open days during Portsmouth Navy Week in August. The carrier had arrived in Britain in July, and her presence was accompanied by an energetic recruiting drive. 430 British recruits were on board when she returned to Sydney on 8 December.

At the end of January, the airgroup began to work up at sea, but a series of accidents sent them instead to the Jervis Bay airstrip for more practice. The problem of the hook bounce of the Fireflies was remedied, with some small modifications in most aircraft, by the time the *Sydney* went to Korea.

5

ANZAC BROTHERS

The big exercises of 1949 were on the occasion of the visit to Australia of the New Zealand Squadron.

New Zealand had seen difficult times with its navy, including a 'strike' early in 1947 which had lost 200 men to the service, on top of the substantial wartime demobilisations. As New Zealand had as many as 650 pilots in the Royal Navy's Fleet Air Arm, there had been some talk of commissioning a carrier, but greater realism was shown by Rear Admiral Simpson, a former submariner who, when he became Chief of the New Zealand Naval Staff, decided to build up an anti-submarine force. Late in 1947, New Zealand sought six modern frigates from Britain, with a seventh to be used as a survey ship. The six ships were of the Loch class and, renamed for New Zealand lakes, they were handed over as the Eleventh Frigate Flotilla between September 1948 and June 1949. The country voted heavily in favour of peacetime conscription at the start of August 1949. The New Zealand Navy, it was stated, was to be trained to defend trade communications in the South Pacific. The build-up involved inevitable manning problems, alleviated to a great extent by the secondment of officers from the Royal Navy. The frigates came with strings attached, namely that three would be at one month's notice at the discretion of the British commander on the Far East Station, to act in support of Hong Kong. With the arrival of the first four in Auckland, the new ships were put through their paces, and the squadron sank a coal hulk by gunfire to show how it was done. The New Zealand Squadron was reconstituted as from 31 March 1949 and the main force sailed from Auckland on 29 September for exercises and to show the flag (with the *Bellona* as flagship). This, of course, included Australia, where they arrived on 3 October. The *Rotoiti* arrived ahead of the rest, at her best speed, as one of her seamen had a ruptured appendix. In command of the *Rotoiti* was Lieutenant Commander A. C. B. Blanfield, RN, who had earned a DSC in Norway and two bars in Sicily and Italy, plus two mentions in dispatches. Aboard the squadron were 1200 men of the regular navy, including 65 Royal Marines and 103 volunteer reservists. The New

Zealanders were pleased with the exchange rate of £123 Australian to £100 New Zealand.

On 5 October, Admiral Farncomb was succeeded in command of the Australian fleet by Rear Admiral John Eccles, CBE, RN. He had commanded the *Indomitable* in the British Pacific Fleet and had seen action in the latter months of the war against Japan. The appointment of an officer of his status in Australia at this time was a sign of the importance to the Royal Navy of the Australian experience with its new aircraft carrier.

The fleet that sailed from Sydney on 7 October comprised:

Australian Fleet
Aircraft Carrier	*Sydney*
Cruiser	*Australia*
Destroyers	*Bataan*
	Warramunga
Frigate	*Shoalhaven*

New Zealand Fleet
Cruiser	*Bellona*
Frigates	*Kaniere*
	Pukaki
	Rotoiti
	Taupo
	Tutira

Initially they went to Jervis Bay, where the fleet sailing race was held. The Australians observed that, unlike the practice in their own service, New Zealand ratings were allowed to relax on the quarter deck of the cruiser, and there was a daily rum issue. Australia, also, had never raised a corps of Royal Marines such as were carried aboard the New Zealand cruiser.

Exercises occupied a week of day and night attacks. Operations on 13 October were pretty typical, with Sea Furies and Fireflies (divided for attack and defence of the Fleet) engaged in dogfights 1000 feet overhead. Rocket assistance was used with some of the aircraft for take-off, and the results were spectacular, as they were in the air using only half the length of the deck, accelerating from 20 to 100 knots in four seconds with a tremendous 'whoosh'.

The big ships formed in line, with the *Bellona* leading the *Sydney* and *Australia*, as all three towed splash targets for the aircraft. The Fireflies attacked these with rockets after a steep dive. Then there was anti-aircraft practice against a target towed by a Beaufort bomber. The guns crews of the *Sydney* carried off the laurel with a direct hit.

At night, the fleet split into two, with the *Sydney* and the frigates representing a convoy to be attacked by the two cruisers and destroyers. After they had been discovered by radar, the ships of the convoy scattered, but those nearest the coast were subject to close-range attack by a 'coastal force' of two air-sea rescue launches.

After her return to Jervis Bay, the *Bellona* suffered an explosion in one of her turrets which left an officer and three men with very bad burns. They had been clearing up, and a fuse had been ignited, setting off cordite.*

All except the *Bataan* went on to Melbourne, arriving on 20 October. As the main part of the fleet came up the Bay, 12 aircraft from the *Sydney* flew over the city and Geelong, landing back on board before the carrier came through the Heads. It was well and truly Navy Week in Melbourne. It was unusual to have the commanders of two navies in port at the same time. On the Australian side was Rear Admiral Eccles, and on the New Zealand, Captain D. Hammersley Johnston, RN. On Trafalgar Day, 16 officers and 300 men from Flinders Naval Depot, complete with massed bands, marched through the city in bright sunshine, and the Governor and the Premier made speeches to the effect that the glory of the day would never be forgotten while there was a British navy. A few days later there was a ceremonial changeover of command of the *Australia* from Captain Henry Burrell to Captain George Oldham. On 28 October, the crews of the ten warships marched through the city. The ships were open for inspection on two weekends and on Melbourne Cup Day when 11 450 people went over the new aircraft carrier, pausing at that certain time in the afternoon to see if their money was on the right horse. All men in naval uniform were given free entry at Flemington. It might be remarked that, although they didn't pay to get in, many paid to get out. The next day the fleet returned to sea.

Off Jervis Bay they tangled with RAAF Mosquitoes and Mustangs. On 10 November 1949 the *Australia* fired her 8 inch guns and was worked up to 25 knots, just to prove that she was not done yet. The New Zealand fleet moved on to Hobart, where they arrived on 14 November.

A return match was played in New Zealand waters in 1950. The Australian contingent worked up together in the Tasman Sea before arriving in Wellington on 25 February. The men of the *Sydney* took the opportunity to admire the spectacular scenery of Milford Sound on the way. The submarine *Telemachus* had been active in 'sinking' the ships as they sailed. At their berths alongside Aotea Quay, they were visited by many people, including 8000 who managed to get aboard the *Sydney*.

---

* In this I have followed the press reports. R. J. McDougall in *New Zealand Naval Vessels*, p. 33 states that a partly fired gun charge flashed back when the breech was opened for reloading, injuring six.

Admiral Eccles was prepared to be defensive about the *Sydney*'s lack of speed, but it was not necessary, and he went on to say, 'We are rather alone in this part of the world, and the more Australia and New Zealand can get together and study each other's problems in sea protection the better'.[1] Great hospitality was shown by the people of Wellington, returned in small measure by an 'at home' aboard the aircraft carrier, whose hangars were decorated with flowers and patriotic New Zealand scenes for several hundred guests.

As for the sailors, they said Wellington was 'far better than Sydney. We had a bonza time. There was more hospitality offering than we could use. It's the best city we've visited.'[2]

On 3 March they sailed to do battle with the New Zealand fleet sailing from Akaroa. Ships involved were:

Australian Fleet
Aircraft carrier	*Sydney*
Cruiser	*Australia*
Destroyers	*Bataan*
	Warramunga
Frigate	*Murchison*

New Zealand Fleet
Cruiser	*Bellona*
Frigates	*Rotoiti*
	Pukaki
	Tutira
	Taupo

British Ship
Submarine	*Telemachus*

It turned out to be a one-sided encounter, as the entire New Zealand fleet, except for one frigate, was declared 'sunk' by the *Sydney*'s Sea Furies, which came in with 'screaming power dives at nearly 500 miles an hour'.[3]

Radar-equipped Fireflies took off at 2 p.m. and located the New Zealand fleet east of Cook Strait. Two Fireflies then shadowed them until the main force arrived. On the other side, the Australian force was reported by radio by the submarine, and land-based aircraft were called in to the attack.

That night there was a surface attack by the New Zealand fleet on an Australian 'convoy'.

Operations went well, and on 5 March eleven ships anchored together at Akaroa as Task Force 75. In the bay there was a sailing race. The Australians, betting four to one against the New Zealanders, lost their money when the

Bellona's whaleboat beat that of the *Murchison* by 20 seconds, with the *Sydney*'s boat third. Some 2000 Christchurch people managed to visit the carrier during her visit.

Next day it was action again as 18 aircraft were launched at ten-second intervals, to attack sites on the Banks Peninsula and inland Canterbury. Observers of the afternoon operations noted that, when an engine failed, the aircraft would have its wings folded and would be wheeled aside in less than two minutes, so that the remaining five could take off. On 7 March the New Zealand Defence Minister, Thomas Macdonald, boarded the *Bellona* to watch the anti-submarine work. As each practice torpedo was worth £4000, they had all to be recovered. The squid mortar of the frigates was particularly effective, and the *Telemachus* was twice sunk by the *Rotoiti*.

The next phase was a convoy action. The *Australia* was designated as a valuable ship being convoyed by the Australian fleet from Napier to Akaroa, and the New Zealand fleet was to attack as best it could. At a crucial point in the action, the Sea Furies were returning short of fuel when they were attacked out of the sunset by RNZAF Mosquitoes. Towards midnight, the frigates made a surface attack, and three were sunk by the torpedoes of the *Warramunga*.

Anti-submarine exercises were next, lent a little bit of a twist by Lieutenant Commander L. E. Herrick of the *Pukaki* having actually commanded a submarine during the war. The commander of the *Telemachus* was Lieutenant O. Lascelles, DSC. Thus there was experience on both sides in their battle of wits. At the end of this exercise, the ships dispersed briefly for port visits, and Lascelles got out his rod and line.

They sailed as a convoy exercise on 12 March but the weather deteriorated as they went north and the submarine only got in one attack, at the mouth of the Hauraki Gulf, as the fleet was arriving in Auckland three days later. It was a grand fleet entry, and most ships moored alongside at Devonport Dockyard. Here they were visited by 15 000 people over the weekend. They were reported as forcing their way towards the *Sydney*'s gangway 'with a determination worthy of All Black forwards'. When aboard, 'Youth gazed with awe at the trim Sea Fury aircraft, the most efficient naval fighter and the fastest piston-engined aircraft in the world'.[4]

The fleet sailed again on 21 March under grey skies, dividing again into rival national forces, but not before the *Telemachus* had tried her luck again. The *Taupo* did not participate in this phase, as she had to prepare for deployment to the Mediterranean. The Australian fleet moved in for a landing to destroy a radar station at Paroa Bay on the Russell Peninsula in the Bay of Islands, defended by the Royal Marines. A Sunderland flying boat shadowing the Australian fleet was shot down, but Mosquitoes attacked in the afternoon. It was not to be New Zealand's day, as their whole fleet was sunk by surface forces in the evening. Owing to a fault with the arrestor wires, the *Sydney's* aircraft were temporarily diverted to Onerahi airport, Whangarei, where one was damaged on landing.

As a finale, the *Bellona* and the destroyers staged a torpedo attack on the *Australia*, followed by full-calibre firings from the New Zealand cruiser against a target towed by the *Rotoiti*. The anti-aircraft guns were exercised against a sleeve target towed at 10 000 feet by a Harvard.

A full regatta of pulling and sailing races was held in the Bay of Islands on 25 and 26 March, and as 'cock of the fleet', the *Bellona* hoisted a Kiwi flag. 'I am a little tired of handing cups to the New Zealand Navy,' Vice-Admiral Eccles remarked. 'I hope the New Zealand squadron will visit us next year when we can take all the trophies back'.[5] To guard against the Australians taking the trophies home prematurely, the crew of the New Zealand flagship rigged plenty of lights and had fire hoses ready at night.

The Australian ships sailed for home on 31 March, and the Sea Furies put on a farewell display of spectacular aerobatics over Auckland. On 4 April the *Sydney* recorded her 2000th deck landing. The figure had been achieved in a very short time, and without serious accident. The crew also noted that they had been able to land aircraft at a rate of one every 23 seconds, which was not bad. It had been a good time. The Australian and New Zealand navies were shown to be at the top of their form, and ready for any enemy whoever and wherever he was to be found.

For the New Zealand navy, this period was one of a slow building of confidence. The wartime works extending and enhancing the dockyard complex at Auckland had been completed in 1947, and the powerful wireless station at Waiouru, which had handled the bulk of the traffic for the British Pacific Fleet, was recommissioned as HMNZS *Irirangi* in 1951.

The acquisition of a seventh frigate (originally to be named *Omapere*) for New Zealand was served by the transfer of HMAS *Lachlan*, which had just passed into reserve. She was handed over on a three-year loan on 1 June 1949 and embarked on a program of surveys in November. An exchange with ships of the British Mediterranean Fleet was made by two New Zealand frigates between April and October 1950. The exchange was useful for experience and training, and harked back to prewar practices. It was no longer felt necessary for Australia.

A reminder of Australia's other ally came with the first official celebrations of the anniversary of the Battle of the Coral Sea, in May 1950. The guest of honour was Admiral Arthur Radford, Commander in Chief of the U.S.

Pacific Fleet, who commenced talks with Mr Francis, the Navy Minister, in Canberra. He was assured that the Australian government had four major proposals to strengthen the fleet:

1. The second aircraft carrier, *Melbourne*, would enter service in 1952.
2. Senior British Fleet Air Arm instructors would be brought out to assist with developments.
3. The *Sydney* would be modernised to handle more advanced aircraft.
4. Extensive overseas training would be provided for Australian pilots.

Radford took the salute at a march in Melbourne, attended a ball in Sydney, and left on a positive note, saying that he would be glad to have good aircraft carriers like the *Sydney* attached to his fleet in time of war. It was not correct to say she was too slow. 'It is a very satisfactory light aircraft carrier,' he said. 'I have nothing but compliments for the way HMAS *Sydney* is going.'[6]

6

THE SHADOW OF NEW WAR

In 1945 everyone had breathed such a sigh of relief that the war was over and the enemy defeated. Yet the postwar world was to be far from peaceful as old scores were settled with several colonial masters and triumphant communism seemed to be moving ever forward. The red flag had flown in Paris in November 1947, the communists had taken over in Prague in February 1948, and the blockade of Berlin had begun in June of the same year. A feeling was growing in many circles that world communism was pushing forward inexorably, and the successes of Mao Zedong in China were just one aspect of this. The takeover of China, where the new government was proclaimed in October 1949, was significant. It was the stated intention of this government to move on to the armed subjugation of Formosa. Who could tell what might come next? Certainly, the fall of Canton had put Hong Kong only 27 minutes' flight from possible attack. In the event, the British were to be surprised at the continued Chinese tolerance of a free Hong Kong.

It seemed obvious what needed to be done, as Australia committed forces to Malaya to combat the 'Emergency'. Initially they comprised Lincolns from No.1 Squadron, and Dakotas from 38 Squadron. They commenced operations against communist forces in July 1950. British forces were in action earlier than this, with aircraft landed from the *Triumph* attacking guerilla positions in October 1949. New Zealand also supplied a transport flight. The navy was not involved in Malaya at this stage, in view of its commitments in Korea.

It was not all one-sided, of course. Nationalist Chinese seized five British merchant ships early in 1950 on grounds that one master called 'plain old fashioned piracy'.[1] One, the *Ethel Moller*, was retaken on the high seas by HMS *Cossack* as it was sailing with 150 Chinese troops to Taiwan from one of the offshore islands. In the great tradition of her wartime namesake, the destroyer put a boarding party over the side on 12 May 1950. The frigate *Whitesand Bay* was ready to provide an escort back to Hong Kong.

The modern reader should not forget that the people who planned and controlled the armed forces in the late 1940s did not think of themselves, as people had in the 1920s, as living in a 'postwar' period. If anything, they were planning for a bigger war still to come, a war in which militant communism would have to be confronted and stopped by armed force. Just when this would happen was anyone's guess, but it seemed likely to be sooner rather than later. In Asia, the communist push seemed to be a direct result of the rebuffs suffered in Europe. The Soviets could not advance there, so they would try advancing in the East. How the next war would be fought also seemed unclear, but the use of nuclear weapons seemed to be quite definitely on the agenda, and this put in question the whole rationale of conventional naval training. Would conventional navies be needed at all, or would it all be over with a series of big bangs?

As usual, the theorists were wrong. When a real war came, it was fortunate that the weapons which lay readily to hand were basically those of the last war, proved and tempered.

The last U.S. combat troops left South Korea in June 1949 and it was seen in some quarters as a significant action, so, on 25 June 1950, the armies of North Korea invaded South Korea and, sweeping all resistance before them soon occupied almost the whole peninsula. By a quirk of the absence of the Soviet representative at the Security Council, the United Nations was able to authorise the use of force for the defence of South Korea, under the overall command of the United States. Australia had Mustangs of 77 Squadron stationed in Japan, these and two ships on occupation duties, *Bataan* and *Shoalhaven*, were immediately offered to the United Nations. As early as 2 July the Mustangs were escorting American B29 Superfortresses and strafing the North Korean tank columns. It was only fortune that found two RAN vessels on station at the one time as one was intended to relieve the other. For operational purposes they were attached to a British Commonwealth fleet, which by early July comprised one aircraft carrier, two cruisers, eight destroyers and six frigates. The United States Seventh Fleet had similar numbers of warships in Japan, but both forces were to be expanded, the American, very much so.

The circumstances of the rapid communist advance, the defence of the Pusan perimeter and the subsequent daring and successful amphibious attack by UN forces at Inchon vindicated carrier-based air power. It was only the carriers that could supply tactical aircraft over the battlefield at a vital time. They were mobile, and they could not be overrun by an enemy army. In fact, it was on the carriers alone that success in the early part of the war turned. Also justified was the U.S. Navy's decision to retain one of its big carriers on deployment in Asian waters.

As things progressed, the Australian ships served with both the Commonwealth and the U.S. fleets, on both sides of the peninsula. The *Bataan* joined a force of U.S. ships comprising the light cruiser *Juneau* and destroyers *Higbee*, *James E. Kyes* and *Collett*, that gave gunfire support to the vital landings at Pohang on 18 July. This was the real thing. It was like the Second World War in the Pacific over again, the old allies together. Only the enemy was different. *Bataan* was back with the *Belfast* firing on Haeju Man on 1 August. *Bataan* and *Warramunga* both formed part of the screen of the *Triumph* during the Inchon operation.

New Zealand had no warships on the spot, but sent the *Pukaki* and *Tutira* as soon as they could be made ready. They arrived in Korean waters early in August 1950. Three Canadian destroyers joined at the same time, *Athabaskan*, *Cayuga* and *Sioux*.

In every case, versatility was required. As the Canadians noted, individual ships were called upon to perform a wide range of tasks. There was no room for anyone who was too specialised. In the case of destroyers, escort against possible air or submarine attack (the scenario which was to occupy more and more of the future of Western navies) was the least of it. Fast supply, monitoring local shipping, gunnery support or interdiction, engaging batteries of guns up to 105 mm at close range and in difficult circumstances, evacuations from dark and badly charted rivers; all required multiple skills. They could be dangerous and all were undertaken with skill and willingness.

In view of the international crisis, a form of conscription was introduced in Australia, as National Service Training, approved by Cabinet in July 1950. An increase in RAN manpower was authorised, from 14 753 to 15 173, and the WRANS was reinstituted with a strength of 300. (It had been disbanded in 1947.) It was stated that 'The Menzies government has recognised that a satisfactory rate of expansion in war is not possible without trained reserves'.[2] It had, in fact, been an election promise based on the widespread perception of continual communist aggression. The plan was that the navy would take 1000 conscripts per year, each man serving 124 days' continuous training, with a further liability of 13 days per year for four years. The first National Servicemen began training with the navy on 30 July 1951. For new recruits to the permanent forces, the term of initial service was reduced from twelve years to six, and for ex-naval personnel, two years.

In addition, training for the RANR resumed on 1 January 1950, under extra funding provided in June 1949. A reserve force whose strength was initially set at 450 officers and 3550 ratings would train on a part-time basis, as well as serving 13 days' continuous service afloat or in a training establishment each year. By 1953, two frigates and six minesweepers, a not insubstantial force, were to be devoted to training duties.

Similar measures applied in New Zealand, where the WRNZNS had been reinstituted in April 1947 to offset a manning shortage, and then made permanent in 1949. The Royal Navy had made the WRNS permanent in the same year.

As early as December 1950, the Admiralty requested the *Sydney* for use in Korea, but Admiral Sir John Collins declined, both because of the upcoming 1951 Commonwealth Jubilee celebrations, and the associated international exercises, and also because of manpower considerations. He felt it would strain the navy too much in the circumstances to have the carrier on a war footing as well as two destroyers.

In April 1951 the request was repeated, with rather more force, but the government was most reluctant to be drawn into what could turn out to be the less relevant part of a full-scale global war, in which its main forces might be required nearer to home. The lessons of the Second World War were still in mind, namely the difficulty that had been experienced then in extricating Australian forces from British command in other theatres. Indeed, the more powerful elements of the British and U.S. fleets were even now being retained in European waters in case of a general conflagration, in which Korea would just be an opening diversion. It was far from peaceful over there. In late 1951 the Royal Navy was single-handedly keeping the Suez Canal open in the face of attacks by terrorists.

For the RAN Commonwealth Jubilee Exercises, a three-phase series of war games absorbed an international fleet in the Tasman Sea over the first three months of 1951. Only the navy was involved, and there was some criticism that, in view of the world crisis, the opportunity had not been taken for a combined-services exercise.

Initially the ships operated in the Jervis Bay area as *Sydney*'s airgroup worked up. Here, she suffered her first fatal accident, on 16 February. Ten aircraft had been out firing rockets on the practice range at Beecroft Head, and the first six had landed when a Firefly, piloted by Lieutenant Robert Smith, made its approach too high. It was signalled to fly through and try again, but the wingtip struck the landing beacon beside the carrier's funnel and the aircraft plunged into the sea. The pilot was killed, but the observer was rescued by the *Tobruk*, and the remaining aircraft were sent to Nowra. The *Sydney* was to have a second casualty on 3 May when the Sea Fury piloted by Lieutenant R.W. Barnett crashed after a malfunction of the rocket take-off gear. The rockets on one side fired on the deck and the others in the air, sending the aircraft straight into the water. In circumstances such as these the men on the destroyer had to be ready to stop by the position of the sunken aircraft, to have all cigarettes extinguished as avgas came to the surface, and to have the whaler in the water ready for an immediate rescue.

As the exercises developed, the ships involved were:

Australian Fleet
Aircraft Carrier	*Sydney*
Cruiser	*Australia*
Destroyer	*Tobruk*
Frigates	*Condamine*
	Culgoa
	Murchison
	Shoalhaven
Tug	*Reserve*

New Zealand Squadron
Cruiser	*Bellona*
Frigate	*Taupo*

Pakistani Squadron
Frigates	*Sind*
	Shamsher

British Ships
Submarines	*Tactician*
	Telemachus

The Dutch destroyer *Banckert* engaged in anti-submarine exercises with Australian frigates off Sydney at the start of the operation. She 'sank' the *Tactician* off Sydney very convincingly. The New Zealand squadron was under the command of Captain J. H. Ruck Keene, OBE, DSC. Admiral Eccles was in command overall.

There was practice in anti-submarine, surface and anti-aircraft exercises, replenishing at sea, night exercises and general seamanship. Just once, a Firefly was flown off in the pre-dawn darkness. As the fleet sailed to Hobart, between 24 and 27 February, Operation Hostile was held. Victoria was declared to be Blueland, and New South Wales and Tasmania, Whiteland, and the object was to get the fleet through to Storm Bay intact, against attacks by Blueland's two submarines, and its powerful surface raider, the cruiser *Australia*. The *Sydney* had extra aircraft aboard, in preparation for Korea, and they struck successfully at the raider, which was then intercepted in a night surface action by the *Bellona* and *Tobruk* and forced to withdraw. The *Bellona* was towing a target for air attack when she was hit aft by a practice rocket from one of *Sydney*'s aircraft on 26 February, and one of her boats was badly damaged. This boat was on the quarterdeck for special attention in preparation for the regatta, and there was talk of malice in the accident until it was discovered that the pilot whose aircraft caused all the fuss was a New Zealander.

The New Zealand frigate *Hawea* joined the force before it arrived at Hobart on 28 February. The submarines lurked in Storm Bay for a last attack and they were harried by Sea Furies and the *Shoalhaven*. In the southern capital there was a fleet regatta, where the *Sydney* took back the prize of Cock of the Fleet from the *Bellona*. To some extent, the naval regatta surpassed that year's Hobart Regatta, which had been marred by poor weather and a polio scare. For the first time, there had been no race for trading ketches. The Jubilee Naval Ball was held with square-dancing called by an American expert, and proceeds to the RAN Chapel at Flinders Naval Depot. As the 14 ships lay in the Derwent, old-timers remembered the fleet that had been there during the Royal Visit of the Prince of Wales 50 years before. No aircraft carriers then, just the cruisers with their black hulls and tall yellow funnels. Now, in the 1990s, men who were there in 1951 remember their fleet with pride. The modern Australian and New Zealand navies can boast no warships nearly as impressive.

The fleet sailed again on 5 March, in deteriorating weather which soon became a 80 km/h south-westerly gale. The *Bellona*, and all the frigates acting as the Eleventh Frigate Flotilla, were attacked by the submarines, and as they anchored next day in Storm Bay they were joined by the minesweepers *Gladstone* and *Latrobe*. On 9 March there was an assault by the 60 Marines from the New Zealand cruiser on Port Arthur, for the purpose of releasing prisoners supposedly held in the ruins of the old gaol. (They were a bit late. The last convicts had been removed from Port Arthur in 1877.) The minesweepers cleared the way and the two cruisers joined forces as the *Sydney*'s aircraft blasted the defences and the Marines went ashore at dawn at Safety Cove. The submarines, allocated to the 'Greenland' defence, attacked the *Australia*.

The *Bellona* and her two frigates were withdrawn abruptly on 14 March, as the New Zealand armed forces were required more urgently to deal with a waterfront strike at home. The bitter strike was to last 151 days, from mid-February (a state of emergency was declared on 21 February) to mid-July, with as many as 20 000 people estimated as being out of work. Naval personnel provided crews to keep coastal vessels and trans-Pacific passenger liners sailing, and at least one railway engine was informally commissioned RNZN. The *Taupo* and *Lachlan* landed men at the Greymouth coal mines. Alongside in Wellington, and later in New Plymouth and

Auckland, the *Bellona* was a potent symbol of force, as well as a source of men to help on the wharves. It was a very serious crisis. Prime Minister Sidney Holland called it a 'very determined effort... to overthrow orderly government by force', and went on to express his belief that the country was 'actually at war'.[3]

The remainder of the fleet dispersed for port visits before the last stage of the exercises, for which the Canadian cruiser *Ontario* joined from Brisbane. Unlike the local cruisers, the *Ontario*, as a training ship, had her armament reduced. On the way back to Sydney, from 29 March to 6 April, Operation Convex Six involved running a Blueland convoy through to Sydney against the opposition of Whiteland. The *Sydney*, *Australia* and *Tobruk*, steaming at their best speed in heavy seas from Adelaide, were the main strength of the Blueland force. Whiteland had the *Ontario*, the two submarines and the shore-based naval aircraft. On 30 March the aircraft had a major success. In low cloud off Gabo Island, 16 of them evaded the Combat Air Patrol and made a successful attack on the carrier which would have left her no longer a fighting unit.

The whole series of war games were highly successful, and future years were to see much fruitful exchange between Australia and New Zealand.

The manoeuvres had a darker background. On his return to Australia from the Prime Ministers' Conference, Robert Menzies indicated his grave concerns. At the meeting of the National Security Resources Board he stressed that the industrial base of Australia had to be ready at short notice to move to a wartime footing. To the state premiers he said that, except in its possession of the atomic bomb, the Western world was, in a military sense, relatively weak:'There is therefore infinite danger for all of us, if the Communists are encouraged to misjudge our temper, our resoluteness, or our capacity for national and individual effort.'[4]

To the federal parliament, he elaborated:

> The dangers of war have increased considerably. It is my belief that the state of the world is such that we cannot, and must not, give ourselves more than three years in which to get ready to defend ourselves. Indeed, three years is a liberal estimate... I am not prophesying war. I merely point out that there is an imminent danger of one; and against imminent danger we must be prepared, and in time.[5]

Mr Menzies noted the pattern of world communism, evident even in Australia, but his orotund phrases were met with a sharp riposte from Mr Riordan on the opposition benches. Labor, he said, had laid down a plan whose main thrust should be further advanced:'If the international position is as serious as the Prime Minister has stated, Australia should immediately begin the construction of additional ships.'[6] In particular, there should be two more aircraft carriers, named *Brisbane* and *Adelaide*, to make a squadron of four. There could then be two ships operational, with the other two in reserve or refitting. It would be costly, but the Prime Minister had indicated that in the circumstances it might be necessary. Menzies, who in 1947 had described the Labor rearmament program as 'extremely meagre',[7] was silent at the barb.

The matter of the commitment of the *Sydney* to Korea was finally agreed by Cabinet on 11 May 1951, but the Admiralty was informed that Australia would reserve the right to withdraw her in an emergency. In the circumstances, Australian fears were mollified by the decision that the refit of the *Glory* would take place in Sydney, and she would thus be present should urgent need arise. To some extent the decision had been forced by the need to have the new security treaty with the United States signed. The Minister for the Navy, Josiah Francis, supported the commitment, on the basis of the prestige that would accrue, as only the third nation to send a carrier to the war, as well as the valuable operational experience. The *Sydney* therefore was withdrawn from her winter cruise in order to prepare for war.

Meanwhile, development continued slowly on the Manus base, helped by the 200 Japanese war criminals. In its own way it was Australia's last convict colony. Visiting sailors appreciated the local girls, who went barebreasted, but it was really a very quiet place. Some remarked that the main duty for the GPV on station was to take the 'dead marines' out to sea to give them a decent burial. The sole RAAF presence was a transport service by DC3s from mainland New Guinea. In 1951 it was decided to begin training men from New Guinea as a first step towards the formation of a locally based and manned division of the navy. On 28 August 1951 the *Koala* left Manus with the first recruits. Enlistment was for three years, with a liability for local service only, and a disciplinary code similar to that for the native constabulary. Communication was in pidgin and Lieutenant Commander David Nicholls was in charge.

In view of the limited storage at Manus, it was not always possible to provide full fuel facilities at the base for warships en route to or from Korea. *Sydney* and *Tobruk* went there on their way to Korea in September 1951 in connection with the need for a show of force at Rabaul. The *Australia* and *Anzac* visited the island in May 1952. The northern frigate patrol resumed in mid-1953 with the *Macquarie* first on station.

In view of the international situation, and the possibility of communist influence becoming established in much of mainland East Asia, America urged that a generous peace treaty be signed with Japan. The U.S. navy wanted the big Yokosuka base, and policy makers saw great value in preserving Japan as a friendly bulwark off the hostile Asian shore.

In Australia, cruel memories were very close to the surface, and there were objections. Polls indicated that 63 per cent of people were against such an action, and the Minister for External Affairs, Percy Spender, stated that 'This settlement must be of such a kind as to contain appropriate safeguards against any resurgence of Japanese militarism'.[8] Unless this was so, Australia could not rest easy. The sentiment was echoed by his successor, Richard Casey. The peace treaty was signed at San Francisco on 8 September 1951, but for Australia, there had to be a sweetener — the ANZUS Treaty.

U.S. Secretary of State John Foster Dulles had wanted to develop a security system based on the 'island chain' of East Asia, but realistically, he had to evolve separate treaties, one with Japan, one with the Philippines, and one with Australia and New Zealand. Britain was to be excluded, as the United States feared becoming involved with problems in Hong Kong and Malaya. Policy-makers in Australia also saw it as a lever to ensure a continued American focus on the region, clear of the probable Eurocentric viewpoint of Britain. New Zealand would have far preferred to include Britain, but was persuaded.

The treaty was also important for Australia, in that it forced U.S. policy makers to differentiate seriously between Australian and British interests in the eastern hemisphere. During the war, MacArthur had never done so. The ANZUS Treaty, signed on 1 September 1951, was never to be as strong as NATO, where an attack on one was an attack on all. Its main thrust was in Article IV, in which each signatory admitted 'that an armed attack in the Pacific Area on any one of the Parties would be dangerous to its own peace and safety and declares that it would act to meet the common danger in accordance with its constitutional processes'. Article V defined an attack 'to include an armed attack on the metropolitan territory of any of the Parties, or on the island territories under its jurisdiction in the Pacific or on its armed forces, public vessels or aircraft in the Pacific'.[9]

Despite Australian assurances that existing relationships with Britain would not be affected, the snub implicit in its exclusion from the treaty was much resented, even to the extent that a newly elected Winston Churchill demanded later in 1951 that Britain be included. But an American newspaper, the *Baltimore Sun*, put the situation trenchantly in 1952: 'No citizen of Sydney or Auckland is likely soon to forget the days when those cities lay wide open to a Japanese attack. The Royal Navy could not protect them then, and probably cannot now.'[10]

The Prime Minister, Robert Menzies, was convinced that world war might be no more than three years away. It would be conducted by imperialist communism and timely preparation for Australia meant the cementing of alliances. If, as the United States and Britain wanted, Australian forces were to be deployed for a world war in the Middle East, then there had to be an absolute guarantee of security in the Pacific. For Australia, ANZUS was important, and for the navy, it was supported by the Radford–Collins agreement in the same year. This addressed the matter of maritime surveillance in the Indian and Pacific Oceans, and apportioned responsibility between the USN and the RAN. The Commonwealth grouping, known as ANZAM, was agreed secretly in August 1949. It extended to 177° east, a line through Fiji and the east coast of New Zealand, beyond which U.S. responsibilities began. Admiral Arthur Radford, U.S. Commander in Chief in the Pacific, a vigorous proponent of a strong and flexible navy, had assured Australia that he would commit sufficient forces to the Pacific to ensure the security of Australia and New Zealand. All of this, plus the Commonwealth connection, was an assurance of security far stronger than had been the case before the war.

7

THE LIFE OF THE SHIP

The organisation of the *Sydney* was the largest and most complex ever seen in the Australian navy up to that time. The world of the destroyer was a traditionally cosy place, comprising some ten officers and 250 other ranks. It was especially cosy compared with the world of a heavy cruiser such as the *Australia*. In the cruiser there was a complement of about 700 men, including 48 officers, or more if the Admiral were aboard. All were organised according to their specialities, including executive, engineering, electrical, gunnery, stores, shipwrights, medical, chaplain, etc., the officers ranging from the greenest midshipman to the captain in command. At any one time there would be the usual movement of promotion, and of posting in and out of the ship. There was the occasional RANVR, but for the *Australia* at least, it was notable how few would include the letters RN after their names. There were newcomers and there were veterans. Notable wartime service gave the letters DSC to Captain Oldham, Surgeon Commander Downward, Lieutenant Commanders Crabb and Dovers, Lieutenant Thomson and Senior Commissioned Officer Pick. There were many other decorations.

The Electrical Branch of ratings had been instituted in the navy in January 1948 — an indication of the technological future that was expected of the service. They were not playing battleships any more; it was a world where good electronics would become vitally important. HMAS *Watson*, the naval radar school, it was said, compared for its size more than favourably with any similar institution in the world.

The standard naval reference books of today list in detail the number and nature of each warship's 'sensors'. The postwar world was one in which sophistication in this area was starting to become important, and reliance on the 'mark one eyeball' and the 'voice powered telephone' would not be good enough. Training was therefore as important as keeping up with the technology.

It was traditional and necessary that a big ship be run with more formality than a small one. Aboard the cruiser, any stoker who dared put his boots on the nicely scrubbed timber of the deck was in for an instant reprimand.

Stokers were never warned when the guns were to be fired. The unwary could find themselves covered in a light ash of asbestos insulation.

But for those who thought that the *Australia* was big, the aircraft carrier would have been an eye-opener. For a start, she embarked a complement of as many as 1300 men, more if there were extra aircraft. There was a distinction between the flyers and the other men of the ship. When fully equipped as a fighting unit, as she was when she went to Korea, the *Sydney* carried 95 officers for the ship itself and 55 officers for the airgroup.

In her earlier years there was a very strong component of Royal Navy personnel on secondment for this new Australian venture, but they became fewer as time went on. In 1950 for instance, three out of four commanders, five of seven lieutenant commanders, and 17 out of 42 lieutenants of the executive branch were RN. Staffing in other branches was similar. Just one year later it was quite different, with the British to a great extent replaced by Australians. It was the navy's intention to have the manning of the fleet up to 85 per cent Australian by 1952.

Australia had made service in its navy specially attractive to men with experience in the British parent service. Many of the Royal Navy sailors had signed on for six years. Of the *Sydney's* original complement, 150 Australians brought back English brides. The first 25 pilots of the 20th Carrier Air Group comprised 14 RN and 11 RAN flyers. Every effort was made to integrate the British and Australian crew members aboard the ship, to ensure teamwork of the highest standard.

New Zealand, which had built its postwar navy up from a very small base, had proportionally many more Royal Navy personnel serving aboard its warships. The *Bellona*, even in 1952, counted 12 of its 29 officers RN, and the Royal Naval component extended to all commissioned warships. The only exception was aboard the survey vessel *Lachlan*, which counted one RN and two RAN among its nine officers.

The accommodation aboard the aircraft carrier was arranged on relatively traditional lines, with the officers aft and the men forward. Nearest the bow, on the gallery deck, forward of the hangar were the CPOs' and POs' messes, followed by the aircrews' messes. Below the forward part of the hangar were the seamen's and stokers' messes. Aft of the hangar on the main deck, were the wardrooms and the gun room, followed further aft by the junior officers' quarters, and above them the admiral's cabin, the captain's cabin and the senior officers' accommodation.

If you were a stoker and the ship was undergoing strenuous wartime operations, you might not see the daylight for days on end. Dave Oliver remembers putting on his shorts and sandals to go out on deck for the first time in two weeks, and being amazed to discover that it was snowing. He remembers the grinding pace of the work — four hours on and eight hours off, continuously for 17 days. Everyone was dead tired at the end of it, scarcely caring if they were dead or alive, but they could indulge in some pretty solid drinking in the brief times when they got on shore. The results of this were often retold with relish.

A recruiting advertisement for the navy stated:

> THE NAVY is a man's life, offering action, adventure and travel. Each warship is a community of trained specialists requiring many different types of recruit. Entry is by selection . . .[1]

They were young men and they were enthusiastic. The ship and all the paraphernalia of modern warfare gave good cause for their confidence.

The *Sydney* had to have modern radar gear for her own use and those of her aircraft. As her gunnery fit was for close range only, her defence against enemy air attack would mainly be her own aircraft, and they needed sophisticated control. There was surface search and height finding, air and surface search, and long-range air search. The long-range radar was good for over 300 kilometres. There was also short range precision radar for air traffic control, radio telephones and a homing beacon. It was admitted that there could be a problem in the swamping of the radar by a concentrated attack by a large number of aircraft. It was one reason why Australia was anxious to be able to operate two carriers together. The plot on a single ship was limited to a maximum of 12 targets, and errors of six to eight kilometres were common. To control the defence against a serious attack would be difficult. Fortunately, it never happened, but it was still prudent to count two light fleet carriers as equivalent to one full-size fleet carrier.

The *Australia*, which was specially fitted for fighter direction, would be able to assist a single Australian carrier in war, as she had a similar fit without the specialised carrier gear, but with other radar for gunnery. The *Sydney* also had a type 72 beacon to assist the return of its aircraft. It was good only to a distance of 80 kilometres, but it could be used in conditions of radio silence. The gunnery of escorting ships, including the cruiser, would also be invaluable in defence against a major air attack.

As for the *Sydney*, flying was the sum total of the ship's purpose. The arrangements for the aircraft provided problems not present on other ships. Avgas, for one, was dangerous and had to be handled properly. Refuelling points were at the edges of the flight deck and the aircraft were refuelled by long hoses. Portable wind barriers could be used if necessary.

The deck, when aircraft were operating, would always be a hive of controlled activity, and the sound of their operation would dominate every other sound throughout the entire ship. Hand signals would be used for every action on the deck because of the aircraft noise. The captain, in manoeuvring, had to be careful of the heel of the ship while aircraft were taxiing on the deck, and the handler had to account for the aircraft and make changes if there were any problems. Everyone had to know what they were doing as a single mistake could cost lives. It was a relatively small space for the number of aircraft, and their dangerous airscrews, and a kind of fatigue could be built up just from the noise. Aviation is very unforgiving of those who make mistakes, and it is most unforgiving aboard an aircraft carrier. As Jacky Fisher was wont to say, 80 years before, what is required are three things, 'Efficiency, *Efficiency* and EFFICIENCY!'

Among the pilots there was a constant striving for excellence. For them, flying was like an addiction. It was glamorous and competitive, and the energy could be contagious throughout the whole ship, especially if things were going well and a large number of sorties was being flown. There was no way that things could be allowed to fall into a repetitive regime. That way there would be accidents.

The men had to be 'peaked' almost continually. 'You can't mix play and carrier flying. Carrier flying is a thing that has to be done all the time. It is too easy to lose touch, and mistakes are generally costly, if not in lives, at least in materials.'[2] As an American flyer stated, with characteristic bluntness, 'If you have a couple of ordinary days you are out of there'.[3] In Korea, the days when no flying took place were valuable to allow the men to unwind, to be ready for next time.

Normally, aircraft would be spotted on the deck for a rolling take-off, with the more junior officers first and the senior last. From the time the starter cartridges were fired there would be a continual roaring of engines to drown speech. Time for last minute checks —

trim — elevator: neutral, rudder: fully left,
throttle — max R.P.M.,
fuel - check contents of main tank; booster pump on,
wings - spread and locked,
flaps - take-off (maximum lift on the catapult)
tailwheel - locked

...and all the rest. The pilot would apply full throttle, release the brakes and adjust his controls for engine torque and, with luck, soar off the deck like a very noisy bird. As each plane took off it would 'jink' to starboard and then continue on its set course. This prevented the next aircraft from being affected by its propeller turbulence. If an aircraft was being launched from the catapult the pilot had to be careful that he did not pull the stick back at the moment the impact was applied. It was all controlled routine as the pilot fired the starter cartridge, did his radio and cockpit check while the deck handlers made sure the aircraft was tight against the launching strop. The flight deck officer would hold up his green flag; the pilot would apply full throttle and drop his right hand. Down would come the green flag, the engineer would put his foot on the pedal and the catapult would fire. The aircraft was secured by the tail also, and this rope had to break before the aircraft could move. This ensured that maximum thrust was being applied at the right moment.

The weight of an aircraft, especially one fully armed for combat operations, could be critical. A Sea Fury needed to have a speed on the clock of 105 knots as it left the catapult. Of this, the ship could supply 24 knots and the catapult 66 knots. There had, therefore, to be another 15 knots of wind over the deck, or the aircraft might not get off. The stronger the wind, the greater take-off weight that could be achieved. This applied especially when an aircraft was carrying a heavy load of bombs. When the ship was in Korea, Noel Knappstein remembers, 'there was always plenty of wind'. In still air, rocket-assisted take-off was required. Whatever its advantages, this method was not liked by the flyers and was rarely used aboard *Sydney*. Pilots were nervous about going off 'in a shower of smoke and sparks'.[4] The catapult, likewise, was 'like being kicked up the backside by an elephant',[5] and the usual method was just to take off with the aircraft's own power, after a run along the deck. In a free take-off, the assembled aircraft would be spotted on the deck aft, with their engines running, and the junior officers, who went first, would have to become airborne with the shortest run. The increased skill required was one of the penalties that went with being junior, and a lot of stories were told at various times of aircraft vanishing in the direction of the water as they left the deck, and then gallantly clawing their way up again.

Allan Benson Heyward, RN, suffered the indignity of having his engine stall just as the catapult fired. Luckily there was not a major accident. His Firefly was carrying two depth charges at the time which could have made a nasty dint in the bottom of the ship if the aircraft had gone over the side.

In the air, each group would form up on its leader and fly off on its assigned mission. Aboard the ship, the Aircraft Control Room would be responsible for the planes in the air, to know where they were, if they had enough fuel or whether they should be brought in hurriedly.

On their return, the aircraft would orbit the ship, flying down the port side at a height of about 400 feet and approaching from the stern. This, rather than a straight approach, would allow the pilot to observe the landing of the previous aircraft, and also the signals of the batsman, without his engine cowling getting in the way. It would also allow the gunnery defences of the ship to sort out friendly aircraft from any possible enemy.

The deck control landing officer stood on the port side, right aft. His gestures were given added visibility to the pilot by the use of bats, and he took over when the approach beacon system left off. The batsman's signals had to be fairly basic, but indicated aircraft speed, height, turn rate and a final cutting of the throttle. Although the pilot's own judgment would usually go hand in hand with that of the batsman, there was no final discretion given to the pilot, as there was in the American service. The batsman's signal for 'cut' and 'wave off' were mandatory.

If the batsman held his arms horizontal, that indicated 'roger'. If he held them up in a V, the aircraft was too high, if he held them down it was too low. There were gestures to tell the pilot to add or reduce power, or to raise the attitude of the aircraft relative to the deck. Finally, there would be the gesture of the right bat across the body, which meant 'cut'. The pilot would bring the aircraft's tail down from horizontal to catch the wire. If the ship were pitching in a heavy sea, the timing had to be just right.

The way an aircraft landed would be watched keenly by other pilots. Even the ordinary sailors would gather if they had spare time to watch the flying. They were called 'goofers'. One or two of them would have a camera ready, just in case. There was great rivalry amongst the pilots for the greatest demonstration of skill. A Firefly might come in at 90 knots and a Sea Fury at 100 knots, though the approach speed of a Sea Fury could be as low as 92 knots if the aircraft were light and clean. Then, with luck, the aircraft would take the wire neatly and be brought swiftly to a standstill. If it did not, there was the very expensive sound of it hitting the steel net of the safety barrier — or worse. If the aircraft caught its wheels in the barrier it was known unofficially as the Grand National. Dave Oliver remembers how once when he was watching the flying from the back of the island, an aircraft bounced heavily and seemed to be coming straight at him. 'How I got out of the way I'll never know.' Men at the guns in the sponsons were always wary of suddenly having to share their perches with an errant aircraft or some of its parts. Geoff Litchfield had a close call with his Sea Fury once, as he almost had a 'torque stall', in which the power of the engine would overcome the wind resistance of the aircraft's flying surfaces and flip it over on its side. Some men, like Norman Lee, could boast that they had only once bent an aircraft, and that was on the very first practice landing on a different ship.

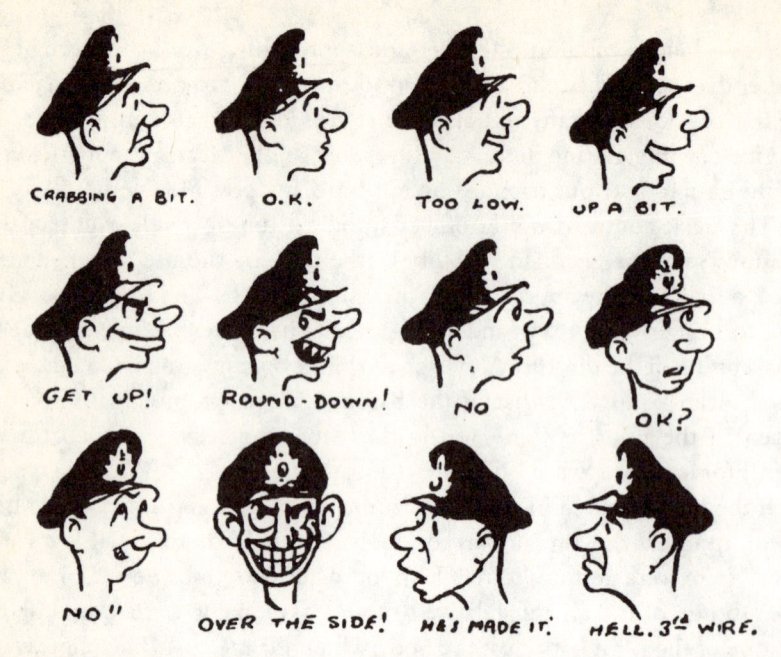

'A NAVAL PILOT WATCHES HIS BEST FRIEND LANDING AN AIRCRAFT ON DECK'
(D. Jones Collection)

With a well worked-up airgroup it would be something to see, with the aircraft coming in like birds one after the other, with the barrier up only for a few seconds between each.

As the aircraft was brought to a halt the hook men raced across to disengage the hook, the arrester wire was reset, the crash barrier was lowered, and the aircraft taxied forward to the deck park. Meanwhile the deck team would be hurrying to fold the wings. Aircraft newly landed were likely to cover the forward lift. In the worst case, the ship would not be able to recover all her aircraft until some were struck below, a process that could slow things down considerably. The lifts could operate to a 36-second cycle, should the handling crews on deck be sufficiently speedy.

For rapid turnaround, the aircraft might then be refuelled and rearmed, and again spotted aft ready for take-off. Even when not in a war zone, 20 hours' flying per month would usually be required to keep each pilot up to standard.

A problem with night operations was the way in which the exhaust flames of the Firefly interfered with the pilot's vision as he orbited the carrier before landing. When the Canadians tried it aboard the *Magnificent*, four aircraft crashed on the first night. The *Sydney* did not indulge in night flying.

Just occasionally there would be a tug strike, and in these circumstances the ship could be manoeuvred in and out of a berth by running the engines of aircraft parked sideways on the flight deck. The sailors on these occasions would be quick to watch the skirts of the girls on the wharf as they were caught in the slipstream. Very artistic!

There were other things, simple things, that needed attention. Every few days the deck had to be cleaned of rubber slicks and grease. In Korea there was also the anti-frost paste from the propellers of the aircraft. Also, the arrester wires had to be checked and replaced regularly, to ensure that there would not be a failure that would cost lives.

Aboard the carrier, as in all professions, there were initials to stand for things which the layman would find completely obscure:

ADDL — aerodrome dummy deck landings
CAP — combat air patrol
CONCAP — convoy CAP
IFF — interrogatory - Friend or Foe?
RAS — replenishment at sea
TARCAP — tactical armed reconnaissance CAP

There were others. The airmen used them like the code of a freemasonry. After starting with ADDLs, a rookie airman would expect to spend some time on the training ship. In the Australian navy, this was to be a major role of the second carrier. There would be practice in formation flying, battle drill, weapons training, navigation across land and sea — all of the skills that would mean the difference between defeat and victory in a real war.

On board ship, there was the need to ensure that everything was done properly, all safety procedures followed — everything by the book. The war had shown that the ship whose procedures and practices were well rehearsed might well survive a crippling attack, and the ship where this was not the case might easily be sunk. Everything meshed together, at all levels of discipline.

Every rank and profession on the ship had its nickname. These are just a few:

Ops	commander (operations)
Wings	commander (air)
Little F	Lieutenant Commander (flying)
Fido	flight deck officer
Badger	flight deck engineer
Bombhead	flight deck ordnance rating

Chockhead	aircraft handler
Grubber	aircraft mechanic
Greenie	flight deck electrician

In the Korean war zone there would also be a detached group of army personnel aboard. They were the Carrier Borne Ground Liaison Section, whose duties would include forward air control and the identification of detail on aerial photographs. The navy referred to them as 'tame pongos', and it was usually found that the natural inter-service rivalry was considerably diminished by men who could appreciate the style and discipline of the other.

The *Sydney* had cafeteria messing, which was seen as a big improvement. For one thing, it made the crew work as a team, instead of as members of messes or flights, which in other ships had been blamed for cliques and fights. The men of the ship took great pride in her cleanliness. There was also a strict rule about the wearing of the dress of the day in the messes, and not just working clothes. Some men set up their own firms to do jobs for the others, such as hairdressing, and there would be the usual gambling and a variety of lurks. There was also the uniform.

> The sight of a matelot putting the traditional seven horizontal creases in his bell-bottoms is not soon to be forgotten.
>
> First the trousers are turned inside out and the distance between each crease is measured carefully. Then the iron is applied lovingly.
>
> There are scores of irons in the *Sydney*.[6]

Very well remembered was Vic 'Jesus' Zammit and his canteen. Jesus was Maltese, and he was reckoned to have made a fortune selling smokes to the sailors. For all that, he was universally praised as a 'good bloke'. The great majority of the men were known to each other by their nicknames. The canteen also sold photos by the official photographers. You could buy a print of your friend's aircraft in an embarrassing position.

On payday the men lined up on the flight deck to give their number and to have their pay put into their upturned cap. They would have to be quick to put the other hand on top, in case the wind caught the notes.

In good weather, the men enjoyed sports on deck — you could enjoy a much greater variety of games than on the deck of a cruiser. A favourite was deck hockey, played six a side with a rope ball. Even with the ship rolling 15 degrees they would race the ball along the deck. 'For sheer speed and daring', observers reckoned, it beat both codes of football.[7] The men could also swim when the ship was in an anchorage away from towns. Also, there would be at least one enthusiast to fly a radio-controlled model aircraft off the deck. Films would be shown in the after lift well with the fire curtains drawn. There was room for 200 men. Favourites of the time were Humphrey Bogart in *The Treasure of the Sierra Madre*, Broderick Crawford in *All the King's Men*, and Bette Davis and Anne Baxter in *All About Eve*. Veterans might have had a few things to say about Gregory Peck as a bomber commander in *Twelve O'clock High*. In Korea, the movie fare was more limited, but film of the bombing runs always went down well, when interspersed with a few cartoons. Men might also be found singing the popular songs of the day. Favourites were 'Night Riders in the Sky', 'If I Knew You Were Comin' I'd of Baked a Cake', 'Come on-a My House' and 'Don't Let the Stars Get in Your Eyes'.

There would also be stories — more elaborate with every extra ale — of daring flyers who played with the limits of performance of their aircraft, or who shocked the bourgeoisie by flying a Sea Fury upside down under the Sydney Harbour Bridge.

As for travel, there would always be a 'buzz' going around about what was to happen next. In foreign ports leave would usually be given, three hours to half the crew in the morning and three hours to the other half in the afternoon. Foreign diseases picked up by the sailors were a serious worry to those concerned about the efficiency of the ship. When faced with a pretty woman, however, most sailors tended to forget the lectures of the medical officer.

After time in port refitting, there would be a need to work up the crew to ensure that flying would be safe and efficient. ADDL would be undertaken at Jervis Bay, with the airstrip marked out to replicate the deck of the carrier. If a pilot had not flown for a certain length of time he would need to requalify, and six successful deck landings were required so that he could re-embark as a fully efficient member of the ship's company. There was the incentive of having extra money paid for flying duties. Then there would be weapons training, such as bombing the long-suffering Beecroft Head. Any variation in the performance of an aircraft had to be noted, to keep it at the peak of efficiency. In one famous report, the pilot filled out the form for his Sea Fury, 'Engine sounds like the clappers of Hell'. In due course, the technicians were able to reply, 'Clappers removed, engine serviceable'.[8] There was work, there was humour, but everything was to the benefit of the whole.

So, with all of this, Australia had a ship to be proud of, and one which was to prove its efficiency in the war in Korea.

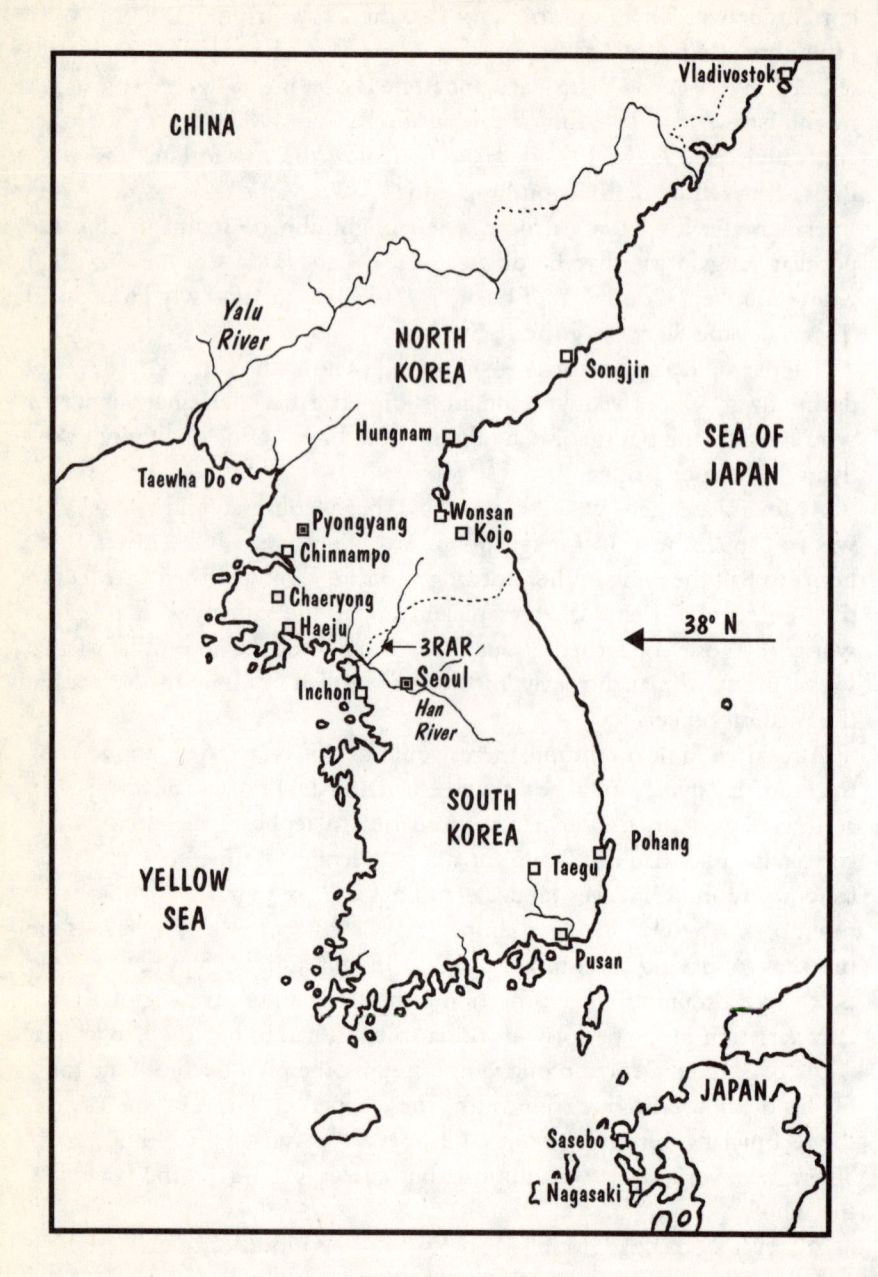

KOREA, SHOWING SOME OF THE PLACES RELEVANT TO THE NAVAL WAR.

8

THE TEST OF BATTLE

The Australian frigates and destroyers had been busy as the fortunes of war surged back and forth along the Korean peninsula. The United Nations command used the weight of amphibious and carrier-borne naval power to full advantage.

The 'Happy Valley', USS *Valley Forge*, was the veteran of the war. With her two squadrons of Panther jets, two squadrons of Corsairs, and a squadron of Skyraiders, she had worked hard. Indeed, by the time she was withdrawn for a rest in mid-November 1950, her airgroup had chalked up 5000 combat sorties with the expenditure of 2000 tons of bombs and rockets. She was back again urgently in December with a new airgroup which, in the next three months, chalked up 2580 sorties, and 1500 tons of explosives.*

As the Chinese forces began to push the Americans back during November 1950, there were dangerous evacuations and other operations. Ships from the navies of the United States, Britain, Australia, Canada, New Zealand, the Netherlands, France, Thailand, South Korea and Colombia all worked together, under the air umbrella provided by the carriers. It is a small curiosity that the conflict in Korea involved more NATO navies than have ever fought together on any other occasion.[1] A South Korean LST, and the *Prasae*, a Thai frigate, were the only major ships to be lost, though several U.S. destroyers suffered crippling damage, with bows blown off by mines. There was work, and hard work too, for all the allied warships in Korea.

The *Bataan* and *Warramunga*, three Canadians, *Athabaskan*, *Cayuga* and *Sioux*, and the American *Forrest Royal* were engaged in an operation of exceptional difficulty in the evacuation of Chinnampo in December 1950. The careful handling of the destroyers in conditions of pitch dark, bone-chilling sleet and dangerous navigation of serpentine channels led to full success in the evacuation of troops and the destruction of all the port facilities

---

★ The airgroup of the *Valley Forge* comprised 30 jet fighters, 28 fighter bombers, 4 electronic countermeasures aircraft, 3 night fighters, 3 airborne early warning aircraft, 2 photographic and 2 night attack aircraft and 2 helicopters.

before the Chinese arrived. It was the stuff of which epics are made. During the bitter Korean winter they continued their patrol work, guarding the strategic offshore islands, engaging shore batteries, and constantly coping with conditions of floating ice, huge tides, poor charts, river silting and the constant threat of floating mines. In June 1951 it was noted that the *Bataan* had steamed the equivalent of a time and a half around the world during her operations in Korean waters. It was stated that she would be returning in July with a new ship's company.'The job goes on even in arctic weather.'[2]

The war began to change, from one of attack and retreat, to a more static phase, when the allied naval effort was directed to the systematic destruction of the enemy's personnel and equipment, and in particular, his communications.'Operation Strangle' began on 1 June 1951 and aimed to obliterate all communications for one degree of latitude behind the Chinese front line with the purpose of bringing the enemy to a more amenable position in the truce negotiations.Yet it was not enough. The dogged and determined enemy repaired overnight the damage of the day, and strengthened his anti-aircraft defences.

Meanwhile in Australia, working up of the carrier airgroups at Nowra had been hampered by 430 mm of rain in June, and while alongside Garden Island on 21 June, some of the *Sydney*'s radar cables were cut by saboteurs. Then, while loading explosives at Shark Island on 4 July she was holed by the boom working horns of the *Kangaroo*. The *Theseus* had suffered similar sabotage before leaving Britain in August 1950. The navy nevertheless laboured to make sure that the *Sydney* had everything possible to enhance efficiency. The aircraft, painted with their black and white stripes for operations with the United Nations forces, flew aboard off Jervis Bay, and with three squadrons aboard instead of the usual two, everyone was determined to show their mettle.Weather permitting, intensive flying training proceeded, first off the New South Wales coast, and later in Hervey Bay. There were accidents, and stress, including the loss of Lieutenant Peter Goldrick's Sea Fury over the side when landing, but the work-up was to prove its worth in the operations to come. *Sydney* was going to live up to her motto,'Thorough and Ready'.

The stripes on the aircraft were adopted because of the doubtful ability of the U.S. aviators aboard the *Valley Forge* to distinguish a British Seafire from a North KoreanYak-9. The two aircraft at first glance had a superficial resemblance. They harked back to the stripes painted on allied aircraft for the invasion of Europe in 1944.

The destroyer *Anzac* departed from Brisbane for Korea on 2 August 1951, and operated with great success on both sides of the peninsula before escorting the *Glory* back to Sydney, arriving on 24 October. Her rapid-fire radar-controlled guns proved their worth against communist communications and as it was the first time this new mark of gun had been tried in action, the British and Australian authorities were looking on with great interest. The 4.5 inch Mk 6 was the new standard for the Royal Navy, and the *Anzac*'s experience was very positive. Fully automatic, with a rate of fire of 25 rounds per minute, they were accurate to a range of ten miles. She had blown up command bunkers at Haeju, destroyed a train near Songjin, supported raids, defended island strongholds, and much else besides. She was good. As someone remarked, she was putting her shells as much as seven miles inland and hitting things the size of a dunny. A story is told of a practice anti-aircraft shoot in company with some American destroyers. The Australian gunnery was hitting the target so readily that the ship was told to go to the end of the line to give the others a chance.

On 31 August 1951, the *Sydney*, under Captain David Harries, and accompanied by the *Tobruk*, left Sydney. The airgroup was under Lieutenant Commander Mike Fell, RN. Both were to earn different grades of the OBE for their work in Korea. The squadron commanders were Lieutenant Commander Jim Bowles RAN (805 – 12 Sea Furies), Lieutenant Commander John 'Apples' Appleby, RN (808 – 12 Sea Furies), and Lieutenant Commander Richard Lunberg, RN (817 – 12 Fireflies). The Commander (Air), Operations Officer, Flight Deck Officer and Landing Signals Officer were all RN. All the crew dressed ship as 12 aircraft from Nowra flew over in salute, and a Wirraway buzzed her deck as she rounded Bradleys Head. The *Tobruk* had been substituted at the last moment for the *Bataan*, which had boiler trouble. Some officers had gone on ahead to observe operations on the RN carriers. In Korean waters she would join Task Force 95, the United Nations Blockade and Escort Force, for operations mainly on the west coast. Here she was to operate as Task Element 95.11, the West Coast Carrier Force. The big American carriers worked in deep water off the east coast, and were considered more secure in that environment. TE 95.11 was usually stationed near the 39th parallel for close support to the western end of the battle line, and for blockade. In this position, near Clifford Island, it was between 70 and 80 miles from the coast, and clear of shallow water.

The west coast role had been taken recently by the *Glory*, working in tandem with a U.S. escort carrier. The USS *Rendova*, with her Corsairs of Marine Fighter Squadron 212, achieved a record for an escort carrier of 64 sorties in a day on 17 November. American ships of this type had a nominal aircraft capacity of 33, but a speed of only 19 knots. The carriers acted in support of Task Element 95.12, divided into four squadrons, which comprised the West Coast Blockade Force, under the command of the captain of the British cruiser. There was also Task Element 95.15, a Korean marine force engaged in the defence of islands.

During her patrol, the *Glory* had lost 20 aircraft, and suffered 115 damaged, of her total of 35 comprising Carrier Air Group 14. Captain K. S. Colquohoun, DSO, said:

> Our aircraft losses were not abnormally high, though they were just about as high as one would care to have. Close support planes in France in the last war got quite a caning too. There is nothing wrong with our Sea Furies and Fireflies. The Chinese go in for light flak in a big way, and when they do there's a lot of lead flying around. Our chaps liked hitting what they aimed at, and came in rather too low. I had to threaten them with a court-martial if any flew below 500 feet.[3]

He noted that the United States Corsairs had armour to protect them, but the British Sea Furies and Fireflies did not. Pilots in Korea noticed that the aim of the flack gunners would tend to be a bit behind the leading aircraft of a strike, but the others would have to watch out. As they put it, by the time No. 4 arrived, the 'gooks' would be well alerted. (The word 'gook' was first used for North Korean soldiers in the same way as 'bogey' was for an unidentified aircraft.) In fact, the last three months of 1951 were the most costly of the whole war in aircraft losses from enemy action among the allied navies. During this time 65 American carrier-borne aircraft were shot down. The *Sydney* was to lose nine in the same period. Her Sea Furies had armour added under the wings to protect the air coolers. They were also modified so that the liferaft was behind the pilot's headrest instead of under the seat. Some were also fitted with oblique cameras.

Operationally, the *Glory* had logged only eight deck landing mishaps, and had scored as high as 1115 consecutive accident-free landings. Going as far as a thousand deck landings without even a burst tyre was a good goal, and a well worked-up airgroup could usually achieve it. Those who remembered the operations of British Pacific Fleet in the war against Japan commented that the standard of flying in Korea was a great deal better. In all, 2892 sorties had been flown from *Glory*, which would have meant a total of well over 6000 flying hours. All of the hardware started to show the strain towards the end of this sort of operation. At one stage she had lost the use of her catapult, and aircraft had been required to take off with rocket assistance. The *Sydney* was also to suffer catapult failure as the stress of her operations mounted. To keep her flyers in trim, the *Glory*'s air group would undergo shore training at Nowra in RAN aircraft while she was under refit.

The *Sydney* lost her first aircraft of the war on her way north, while exercising. At the time of her arrival in Korea, late in September, the major allied warships on station were:

United States ships
  Battleship   *New Jersey*
  Aircraft carriers *Bon Homme Richard*
                  Boxer
                  Essex
  Heavy cruisers *Helena*
                  Los Angeles
                  Toledo
  Escort carriers *Rendova*
                  Sicily

British ships
  Light fleet carrier *Glory*
  Light cruisers *Belfast*
                  Ceylon

Although the U.S. and Commonwealth warships usually operated in their own areas, the experience of joint command in the last war, especially on the Australian side, was of assistance when they came together.

The *Sydney*'s first patrol commenced on 4 October, with 47 sorties next day, including a strike against bridges and communist positions 190 kilometres north-west of Seoul. This was part of the campaign which had commenced late in July to force the front back from the Han River and the approaches to Seoul. Thereafter it was a usual schedule of about 11 days on patrol, of which one would be spent replenishing at sea from vessels such as RFAs *Wave Premier*, *Wave Chief* and *Brown Ranger*, followed by five or six days off. More extended operation was possible, but this rhythm allowed rest and recuperation, as it was found that the crews of some of the British carriers had become very run down as their time continued. Mike Crosley has commented that a patrol of 30 sorties, involving 45 flying hours was 'hard going'.[4]

An early operation was an air and surface attack in the Kojo area on the east coast, in support of advances by Korean forces. Task Group 95.9, under the command of Rear Admiral A. Scott-Moncrieff, RN, comprised:

United States Ships
  Battleship   *New Jersey*

Destroyers	*Colahan*
	Hanson
	Shields
	Renshaw
Australian Ship	
Aircraft Carrier	*Sydney*
British Ships	
Cruiser	*Belfast*
Destroyers	*Concord*
	Comus
	Cossack
Canadian Ship	
Destroyer	*Cayuga*

Sea Furies of 805 and 808 Squadrons, and Fireflies of 817 Squadron, from the *Sydney*, flew continuous sorties, despite low cloud, spotting for the warships as they fired on enemy troop concentrations, and attacked various targets. On the second day, Thursday 11 October, *Sydney* had her busiest day of the war. In all, her aircraft flew 89 sorties in the daylight hours between 6.30 a.m. and 5.00 p.m. for strike, combat air patrol and gunnery spotting for the 'Big Jay' *New Jersey*, the *Belfast* and the two accompanying destroyers. The deck parties would have to have the second strike launched before the first could land, then there was refuelling and rearming to get them away again before the other team arrived back. It was intensive and stressful work, with little time to look over the side at the gunfire of the other ships. Inevitably, there was damage which required aircraft to be struck below and replaced, but by the latter part of the day there were an impressive 31 aircraft aloft at the one time. The fifth strike of the day saw 18 Sea Furies flown off the deck starting at 12.30 p.m. The final strike was by 12 Sea Furies, launched from 3.45 p.m. on 2000 enemy troops caught digging in. Over 200 were killed and an ammunition dump was blown up. This was the debriefing report:

> All pilots reported seeing hundreds and hundreds of enemy troops and estimated hundreds of small store dumps dotted all over the general area of the eastern slopes of the ridge within a mile of DU 013119. Both troops and dumps were repeatedly rocketed and strafed. Troops had been digging in and the trenches were only partly dug with resulting chaos and confusion on the ground with enemy running in all directions. A conservative estimate of the enemy killed or wounded by the Furies is 200.[5]

In view of the status of this day, as a high point in Australian wartime naval aviation, and also as a source of some mythology, this is a detailed listing of events aboard the *Sydney*:

> Event 'A' — 0630 — 0800 hrs
> 7 Sea Furies for strike, 3 for air spotting and 2 for combat air patrol
> Event 'B' — 0745 — 0930 hrs
> 6 Fireflies for strike, 2 for air spotting and 2 Sea Furies for combat air patrol
> Event 'C' — 0915-1115 hrs
> 7 Sea Furies for strike, 3 for air spotting and 2 for combat air patrol
> Event 'D' — 1000 — 1245
> 6 Fireflies for strike, 2 for air spotting and 2 Sea Furies for combat air patrol
> Event 'E' — 1230 — 1430 hrs
> 14 Sea Furies for strike (one group of 8 and one of 6), 2 for air spotting and 2 for combat air patrol
> Event 'F' — 1415 — 1600 hrs
> 9 Fireflies for strike, 2 for air spotting and 2 Sea Furies for combat air patrol
> Event 'G' — 1545 — 1700 hrs
> 12 Sea Furies for strike, 3 for air spotting and 2 for combat air patrol

As each aircraft flying off the deck is counted as a sortie, this makes a total of 92 for the day. Other documents give the total as 89, and if this is true, then three of the sorties listed above did not occur.★

Vice-Admiral Sir Guy Russell, Commander in Chief, Far East, sent a complimentary signal to Captain Harries:

> Your air effort in the last two days has been unprecedented in quantity and high in quality. It has been a magnificent achievement on which I warmly congratulate you. Though it is invidious to particularize, the spotters did a first class job and the *New Jersey*, with the Commanding Officer of the Seventh Fleet embarked, said that they were the best she has had. Eighty-nine sorties in one day is grand batting by any standards, especially in the opening match.[6]

The number of sorties, which came to a total of 147 over two consecutive days, was a record number for an aircraft carrier of her class. The *Glory* had set the previous record, of 84 sorties in a day, in September, and was to take it back early in 1952 with 104 sorties. As the war progressed the record moved steadily upwards, until it reached 123, recorded by the *Ocean* and

---

★ The typed debriefing notes do not give details for combat air patrol, and are themselves marginally inconsistent. The frequently stated assertion that this sortie total was a record for a ship of the type for the whole war is incorrect.

later equalled by the *Glory*. This was substantially better than the level of operations recorded in 1945 during the last days of the war against Japan.

Individual sorties could vary between one and three hours, perhaps two a day for an aircraft in usual operations. Aboard the carrier, an aircraft could be refuelled and rearmed in somewhat over two hours. Thus about 54 sorties per day in five 'events' in the winter months would be the target usually expected. It was certainly the goal that Captain Harries looked for. The big American carriers, with their 81 aircraft, looked for an average operating rate of 100 sorties per day in summer and 85 in winter. The final record went to the *Princeton*, with 184 sorties in a day in June 1953. (In more recent times, the big U.S. carriers have aimed for a more relaxed rate. The *Independence*, with 86 aircraft, averaged 58 sorties a day during her patrol in the Persian Gulf in 1994.)

The various roles of the *Sydney* Carrier Air Group included armed reconnaissance, combat air patrol for various roles including convoy, rescue and naval gunfire support, strike, army co-operation and photo reconnaissance. West coast operations would start with a dawn patrol to the Manchurian border by four Sea Furies, followed by Firefly strikes as required, frequently on railway bridges. A guerilla group led by an American codenamed 'Leopard' would report on results, and joint operations were co-ordinated through Taegu. Each day's proposed activities were plotted for the kind of aircraft and weapons that would be used — delayed bombs for railway bridges, instantaneous for a troop concentration. Each mission would be located on a map — for close support of the army, gunfire spotting for a destroyer, attacks on a railway siding — whatever was required.

Although it never happened, there was always in mind the possible intervention of communist submarines, for which an aircraft carrier would be a fat target. The big ships were always closely escorted and important convoys had to be escorted as well. The only real problems were minefields, and it was usual to keep the big ships well out to sea to avoid them. The main allied fleet base at Sasebo had boom defences and patrol boats.

In fact, although Stalin had submarines at Port Arthur, Vladivostok and Nakhodka, all very near the Korean fighting, he kept the Soviet forces on a very tight leash in Korea.* Soviet jet fighter aircraft, though painted in Chinese markings, were not allowed to operate within 100 kilometres of the battlefront or to fly over the sea. As a result, they were rarely met by carrier aircraft. The allied flyers, not knowing this, were frequently surprised that the carriers were not attacked by MiGs. Only once, in November

---

★ The Soviet Union had a lease of the Port Arthur/Dalian area from 1945 to 1955 and based about 18 submarines there. Mao Zedong saw them as a form of insurance against American aggression.

A Firefly of the 21st carrier air group takes off from the *Sydney* in British waters early in October 1950. (RN)

An Avenger takes off from the *Sydney* in 1949. (A. Porter)

With the wartime stripes painted on the aircraft and extra 40 mm guns on the quarterdeck, the *Sydney* returns from Korea in March 1952. (Herald & Weekly Times)

One of *Vengeance*'s Sea Furies from 809 Squadron flies over the suburbs. (N. Knappstein)

A Sea Fury taking off with rocket assistance at Nowra. The solidity and safety of all that concrete under the wheels was not a luxury to be enjoyed at sea. (A. Porter)

One of *Sydney*'s Fireflies leaves the catapult of the *Vengeance*. (N. Knappstein)

Frigate *Shoalhaven*. (RAN)

The *Unicorn* was vital in Korean operations for the Commonwealth carriers. She provided floating base and repair facilities close to the operating area. She was known to her men as the 'Fighting Ferryboat'. (T. Molland)

The *Tobruk* arrives in Brisbane in November 1956. Her funnel has been modified to keep fumes clear of the bridge. (C. Jones)

The liner *Kanimbla* was retained by the RAN after the war as a troopship.
(D. Robertson Collection)

A Sea Otter tied down on the deck of the *Sydney* for the voyage to Australia. (A. Porter)

Destroyer *Anzac* being refuelled by the *Sydney*. (C. Thwaites)

Noel Knappstein is winched up to 'Uncle Peter' — the name was taken from the recognition number UP-21.
(N. Knappstein)

The *Australia* leaves Brisbane for the last time in May 1954. (H. Jones)

The *Vengeance* arrives in Brisbane in September 1954. (M. Jones)

Left: Noel Knappstein's Sea Fury just hanging onto the side of the ship. (N. Knappstein)

Below: Neatly balanced on its nose, a Sea Fury has won the 'Grand National'. (D. Oliver)

Destroyer *Warramunga*. (C. Thwaites)

The frigate *Murchison* at the city wharves in Brisbane in August 1952. The sky sign advertises the locally preferred brand of 'Nelson's Blood'. (M. Jones)

Above: Danny Buchanan's Firefly has been given the wave off but has lost air speed and is about to crash.
Below: The crashed Firefly has slewed around to face aft. Damage to the other aircraft can be seen. (A. Porter)

Admiral Collins discussed operations with five of the flyers, J. Cavanagh, Jack Salthouse, Noel Knappstein, N. Williams and A. Wright. (N. Knappstein)

The first Firefly starts to move down the deck as the squadron prepares to take off from the *Sydney*. (A. Porter)

Sea Furies being struck below on the after lift. (D. Oliver)

Arming a Sea Fury. (A. Porter)

Three Firefly FR4s, based at Eglinton in Northern Ireland. (D. Oliver)

Aboard the training ship *Illustrious*, the Sea Fury of Lt Leece RN has run into the first barrier. The aircraft is from 736 Squadron, based at Culdrose (N. Knappstein)

Christmas in the stokers' mess. (D. McLeod)

A group of sailors spends Christmas night at Kure. 'What do you think of the girls? Wacko!!' (D. McLeod)

The deck operations officer is ready to give the signal to fire the catapult for this Firefly. (D. Oliver)

Ian 'Tassie' Webster's Sea Fury has hit the round-down and lost one of its wheels, which bowls down the deck ahead of it. (A. Porter)

Flyers of 808 Squadron kitted out and waiting for the scramble in the ready room, January 1953. From left to right they are Arthur Smith, Jack Salthouse, Noel Knappstein, Jack Suriano, Observer Morris and Peter Wyatt. (N. Knappstein)

A 'target card' for an attack on Kojo on 11 October 1951. Each pilot carried one of these to indicate the point to be attacked. In this case it was a supply dump to be hit by bombs and rockets (N. Knappstein)

A bitter winter off Korea; snow falls on the deck of the *Sydney*.
(D. Oliver & D. McLeod)

1952, was there a major clash over the sea, when the aircraft of the *Oriskany* beat off an attack by MiGs. A contemporary Russian joke had it that the Russians found it quite difficult to fight in Korea as they had to do it one-handed. The other hand was needed to pull back the skin around their eyes to make them look like Chinese. It was also fortunate that the mines used by the North Koreans were not entirely state-of-the-art, or more ships would have been lost.

The Fireflies aboard *Sydney* were the Mark 5, rather than the Mark 6, the type configured for an attack rather than an anti-submarine role, and armed with four 20 mm cannon. They were on loan from the Royal Navy. Generally, the Sea Furies tended to work with rockets, and the Fireflies with bombs, in view of the limitations on deck operations imposed by the carrying of the weight of external fuel tanks. The air group commander, Mike Fell, said 'the planes were ideal for the purpose for which they were used — low level bombing, rocketing and shelling. Their air performances were splendid'.[7] After the *Sydney* arrived in the war zone, she was allocated an American HO3S helicopter (the type called the Dragonfly by the British) for plane guard and rescue, in place of her Sea Otter. The ship's company referred to the helicopter as Uncle Peter after its recognition letters, UP 28. With a rescue helicopter in place of a seaplane, flying operations could take place in more stressful wind conditions.

While in Sasebo to store and replenish, there was a warning of the approach of Typhoon Ruth, and the Commonwealth ships put to sea to ride it out. In Captain Harries' case, it was a conscious decision to leave a harbour rated as a safe typhoon anchorage, protected by 2000 foot mountains. He feared that his moorings might not be sufficiently robust, and that the ship might suffer damage from collisions in the crowded harbour. From 14 to 15 October the ship battled the elements, as the wind rose to an estimated force 12 around midnight, and waves up to 45 feet high crashed over the deck. Some 13 aircraft were firmly tied down on the flight deck but the heavy movement of the ship eventually caused lashings to break, undercarriages to collapse, and the aircraft to shift. One was blown over the side, two others fell off the edge into the gun sponsons, and a tractor and a motor dinghy were also washed away. Half of her aircraft were badly damaged. Seven were counted as write-offs and three had to be sent subsequently to the *Unicorn* for major repair. The 32-foot cutter was also smashed. Captain Harries manoeuvred the ship to take the sea at her best angle, fine on the starboard bow. Solid seas coming aboard had entered the ventilation ducts, there were fumes from ruptured fuel tanks on the aircraft, and there had been small fires in electric motors of the ventilation fans, but these were rapidly controlled. There were a number of white faces as the crew remembered the loss by explosion and fire of the *Dasher* in 1943. Down below in the canteen, the crew watched the regular flexing up and down of the floor. *Sydney* came through, and for those who could spare a glance, there was the sight of the two accompanying destroyers battling the mountainous seas. The Canadian vessel, *Sioux*, suffered a bent bulkhead.

Back in action, she was able to celebrate Trafalgar Day with 53 sorties, including a series of strikes on enemy junks in the Yalu estuary. They were believed to be preparing for an attack on Taewha Do. This island was subject to particular attention by the forces of both sides, as its status had been under contention at the truce talks.

In general operations, it was usual for some 38 per cent of aircraft to return to the ship in an unserviceable condition because of enemy fire or some other reason. During the intensive operations in October the figure was as low as 29 per cent, but on a subsequent day, two in every three returned damaged. In all cases, aircraft were fit to fly again by the following morning. The presence of the support carrier *Unicorn* was a major bonus, as she was virtually a floating base, and could provide support much closer to the operational area than the land facilities at Iwakuni. Major M. B. Simpkin was chief of the carrier-borne ground liaison contingent aboard the *Sydney*. He remarked: 'Enemy flak has certainly increased in strength and accuracy. He is using guns ranging from the equivalent of our 3.7's to 12 millimetre. When our planes attacked, everything went up in the air, even bullets from rifles.'[8]

The most popular task was the support of Australian ground forces. Soldiers of 3 RAR were at this time engaged in bitter fighting in the Maryang San area of central Korea to establish a new front — the Jamestown Line. Air support at the right time could save lives on the ground.

A dramatic rescue on 23 October involved the dropping of an inflatable dinghy from a Firefly to a ditched B29 pilot north of the Chinnampo Estuary. The aircraft then patrolled overhead until the man was picked up, in a minefield, by a boat from the *Murchison*. The frigate herself had a taxing patrol, including 60 days in the treacherous Han River, right on the enemy front line, frequently in action at close range with the communist guns. Although the frigates were at very significant risk, the Americans were pressing the British to increase operations there. It was deemed vital in persuading the communist truce negotiators that they had no claim to control of territory south of the 38th parallel.

Although the assignment of the *Murchison* to Korea in place of the *Tobruk* had initially not been looked upon with favour by our allies, her operations between July 1951 and January 1952 were exemplary. The Chinese were liable to prepare an ambush of 3 inch guns and mortars to open fire at ranges as low as 600 yards in the twisting and constricted spaces of the Han River. At the end of September there were some very hot fire-fights, with the four 4 inch guns of the *Murchison* returning better than they got. On one occasion she was hit in the engine room, and had luck not combined with cool handling, she would certainly have been lost. Admiral Dyer, on board during one of these actions, was impressed. Lieutenant Commander Allen Dollard, as the current authority on the hazards of the Han, was showing Lieutenant Commander Turner of the *Rotoiti* the channels when the hottest fight erupted. Turner was also impressed: 'Dollard's handling of his ship and general direction of the armaments was faultless and imperturbable.'[9]

The words of her executive officer may briefly describe the scene: 'There is "X" mounting, the only guns in the ship in action for the moment, trained right aft, its crew working frantically around it, the flashes, the brown cordite smoke, the recoiling breeches.'[10] It was also nice for the frigate to have the *Sydney*'s Sea Furies available at times to spot for mines.

New Zealand, at this time, was represented in Korea by the frigates *Taupo* and *Hawea*. Their single 4 inch guns, mounted forward, rendered them less versatile than the Australian frigates with their twin 4 inch mounted both fore and aft.

Three of *Sydney*'s aircraft were shot down during operations on 25 and 26 October and a fourth was badly damaged. One was during a five-plane raid on a railway tunnel between Chaeryong and Haeju, when a Firefly was shot down in a frozen rice paddy 50 miles behind enemy lines. Sub-Lieutenant Neil MacMillan and CPO Phillip Hancox kept enemy soldiers at bay with the standard issue Owen sub-machine-gun. A protective umbrella was maintained overhead by Sea Furies, but one of these was hit when it swooped to drop a message and had to limp to a friendly airfield. Meteor jet fighters of the RAAF's 77 squadron assisted in the umbrella in case enemy jets should arrive. Meanwhile a shore-based helicopter had had to turn back. At 16.22 the *Sydney*'s helicopter set out to cover the 172-kilometre distance to the rescue. It was accomplished just in time, with all of the aircraft at the absolute limit of their endurance. The helicopter, which had achieved a record speed of 120 knots, was not equipped for night flying but was guided by the flash of machine-gun fire, dropping down swiftly from 4000 feet. Aircrewman Callis Gooding, newly briefed in the use of the Australian Owen gun, shot two of the communist soldiers 15 metres away as they landed. As the helicopter took off again, Sea Furies protected it by laying down a barrage of fire on both sides. Everyone was safe as they landed at Kimpo with the assistance of headlights from trucks. It was an event to be told and retold. The helicopter pilot, CPO Arlene 'Dick' Babbit, was awarded the DSM as well as the U.S. Navy Cross, the only allied serviceman in Korea to receive the awards of two nations for the same action.

Gooding was renowned in the ship as a daredevil. Men were absolutely forbidden to smoke on the flight deck, but as soon as the helicopter was in the air he might be seen, sitting in the open doorway with his legs dangling, his feet in non-regulation socks, and with a cigarette in his mouth. The men on the ship paid him the accolade of saying he was 'as game as Ned Kelly'.

Noel Knappstein also had his Sea Fury shot down on 26 October. He was able to make an emergency landing but the wings of his aircraft were ripped off when he hit a stone wall. Noting that it was never going to fly again, he sold the remains to a Korean peasant for an immensely satisfactory wad of notes to the value of 1000 Wong. He was rescued by a boat from the *Amethyst* and came back to the ship with his pockets literally bulging, but when it was all counted up and translated into Australian currency, it was worth exactly one shilling and ninepence.

The first casualty was Lieutenant Keith Clarkson of 805 Squadron, whose Sea Fury was hit by 20 mm fire while diving to attack a road convoy on 5 November, the first day of a new patrol. Lieutenant Commander Richard Lunberg's Fireflies made determined attacks on railway bridges, particularly those north of Wonto and at Sokton-Ni. The latter bridge fell only after several attacks, owing to difficult flying conditions.

During late October and early November there were operations against junks to prevent the occupation by communist forces of islands in the Yalu Gulf. Among other operations, the *Sydney*'s aircraft and the *Belfast*'s big guns stood ready to support the Canadian destroyer *Athabaskan* as she attempted to draw the Chinese batteries on Sok To. The days were short and the weather far from ideal. The snow, however, always melted on the greater part of the flight deck through the warmth of the hangar below. During this period there was only one accident in 400 deck landings, and aircraft were catapulted off at an average 42-second interval, though once this was got down to 39 seconds.

There was, of course, time to pay attention, at 3 p.m. Melbourne time on 6 November, to whether Morse Code would win the Cup. It was the lucky person with Delta on his ticket, however, who was told he had won the sweep while flying over the bleak Korean countryside.

On 13 November the *Sydney*'s pilots were spotting for the American battleship *New Jersey* as she pounded Kansong and the Chang-San-Got Peninsula on the west coast. The battleship was at the end of her tour of duty and at the top of her form, scoring hits with substantial accuracy with her 16 inch guns.* The contribution of the Australians was warmly commended.

On this day they recorded the 1000th sortie, with attacks on troops in the Han River area, and a strike that blasted three spans of a railway bridge.

In late November 1951, the *Sydney* flew 123 sorties in support of a combined diversionary attack, Operation Atheneum, on Hungnam, an important rail and communications centre on the east coast north of Wonsan. Here, the targets were barracks, industrial plants, stores and rail communications, attacked with the intention of drawing enemy troops away from direct engagement on the front line. Task Group 95.8, again under Rear Admiral Scott-Moncrieff, comprised:

Australian Ships
Aircraft Carrier	*Sydney*
Destroyer	*Tobruk*

British Ships
Cruiser	*Belfast*
Destroyer	*Constance*

United States Ships
Destroyer	*Hyman*
Rocket landing ships	LSMR 401
	LSMR 403
	LSMR 404

Netherlands Ship
Destroyer	*Van Galen*

Canadian Ship
Destroyer	*Sioux*

The naval bombardment was carried out during 20 and 21 November, and it was generally noted by the aviators that the bigger the naval guns, the more accurate was their fire. A barrage of 3600 5 inch rockets was regarded as quite awesome. Those who witnessed such attacks described the sound of the rockets as like that of a giant whip being laid across the land, with their swish and crack. The bigger guns had a deeper sound, while a rocket attack by aircraft was like a gigantic cotton sheet being ripped apart'.[11]

★ The *New Jersey* was to have a long war record. She first fired her big guns in anger in 1944, and for the last time in 1984.

In addition to her operations with the *Sydney*, *Tobruk* attacked bridges and rail junctions south of Songjin. With the mountains and high cliffs of the Korean east coast, the railways were close to the shore and vulnerable. With one salvo she destroyed a supply train, flames shooting 200 feet into the air as the cars exploded. Her star shell had revealed its presence in the night, and although it was racing for the nearest tunnel, it did not have a chance. As one young officer remarked, 'We spread it all over the place'.[12] Rapid and accurate gunfire was a must for the destroyers in Korea. As one Canadian observed, it was 'like snap shooting in a shooting gallery'.[13] Unfortunately for the Australians, the 'Trainbusters Club' was not created until July 1952. HMCS *Crusader* was to be its most honoured member.

As for the airmen, it took technique and nerve to bomb a railway bridge, as a near miss would not be good enough. If a bomb was set with a 25- or 30-second fuse, it could hit its target but bounce off before exploding. With a bridge you needed a direct hit. The Fireflies initially tried dive-bombing, for which they were not really suited. They would come roaring down from 8000 feet and release their two bombs at 3000 feet, but it was difficult to make a direct hit, and in the case of a bridge, this was what was required. The geography of Korea being what it was, there were an awful lot of bridges. It would be a little better to dive diagonally across the bridge and put a bomb into the abutments, but before long the Fireflies adopted an anti-submarine technique in place of dive bombing to succeed, coming in level and low along the length of the bridge in groups of four. They had to be careful in the circumstances of the anti-aircraft fire. Some bridges needed constant attention — smashed by bombs by day and repaired again by night but it was great satisfaction to destroy the spans, and then go on to destroy a diversion bridge on the same river. The enemy trains used to hide in tunnels during the day, and it was also a source of much satisfaction to be able to fly up a cutting and slip a bomb into the tunnel mouth, and then see the continuous smoke that indicated a hit. Bugs Bailey achieved this, though he had to haul his aircraft up again rather swiftly before he hit the hillside. Norman Lee remembers:

> With experience we became quite good at the job and could knock a bridge down with only one aircraft. The bridges were relatively flimsy affairs and would be re-erected by the North Koreans overnight. Next day we would do our rounds and knock them all down again.[14]

During one nine-day patrol of 397 sorties, the count was 14 bridges, 7 junks and 3 tunnels destroyed, and 7 bridges, 54 junks and 5 tunnels damaged, with 1000 enemy casualties. The total would have been higher but for one day of extreme bad weather which limited operations to 22 sorties instead of a more usual 54.

The long series of actions that became known as the Battle of the Islands began on 30 November, when the Communists finally took the island of Taewha Do, south of the Yalu. Allied efforts thereafter went towards the defence of the larger islands further south, and the *Sydney* assisted the other ships in the task, spotting for naval gunnery, destroying enemy shipping and troops, and aiding defensive measures. The air support was much appreciated by the other ships, required to operate within range of enemy artillery in extreme cold and in water sometimes covered with ice. It was unfortunate, when the destroyer *Constance* was holed by communist artillery fire in the defence of Sok To, that the *Sydney* and her aircraft were committed to a convoy escort task a long way out of range.

Two more of the *Sydney*'s pilots lost their lives. They were Sub-Lieutenant Dick Sinclair, on 7 December, and Sub-Lieutenant Ron Coleman, on 2 January, both from 805 Squadron. During the fifth patrol, from 6 to 18 December, the aircraft suffered heavily from enemy flak, with five shot down (Sub-Lieutenants Smith and Sinclair, Lieutenants Oakley and Cooper and Lieutenant Commander Gowles) and 20 damaged. During her whole tour of duty, ten of the *Sydney*'s pilots were shot down, but seven were rescued. The enemy would fire everything they had, from rifles to 12 mm shells, at the aircraft. You could be lucky or unlucky. One of the lucky ones came back with the hole from a bullet right up the padding of his seat, just an inch from his spine.

From the personal point of view, all of the crew knew how Sinclair had been found in his carley float with hideous wounds, and remembered that he left a one-month-old baby back at home. There were 1200 men on deck in light snow with tears in their eyes as he was buried at sea with full honours.

On 13 December the helicopter piloted by Lieutenant Raymond Smith, USN, performed a double rescue. First, Lieutenant Peter Cooper was picked up where he had crashed on land, while patrolling aircraft kept off the enemy, and then Lieutenant Commander Walther Gowles was winched up from where his plane had gone down in the sea.

In the absence of the *Unicorn*, the *Sydney* returned to Kure for aircraft support, and there spent a week over Christmas. It was a good break for everyone, with tinsel and Christmas trees. Everyone received something from Australia, even if it was only a telegram from the family. Christmas on a warship is a day of misrule, when the sailors dress up, some displaying many more medals than they could possibly win in a lifetime. The lower deck remembered the flyers parading through the ship banging drums, singing songs and blowing balloons.

Ashore, it was said that you had to have a pocket bulging with notes just to have the price of a soft drink. Kure, of course, had other attractions, as fighting men are seldom celibate. 'The 110' in Duke Street next to the Cambridge Hotel was a very welcoming establishment with an 'all Native floor show'. 'Our ladies are the best for your entertainment,' they said. Mike Crosley, who was in Kure on the *Ocean*, described it like this: 'It is really like "Little America" and has all the usual gaudiness of an American film set in a goldrush town — flash cars and popsies.'[15] The American shore patrols, however, were hard faced men who took no nonsense.

The ship was back at sea on 27 December for her sixth patrol. Many of the crew, who had not seen snow before this cruise, were allowed to indulge in a snowball fight on the flight deck as the ship sailed off station in early January. It must be remembered that *Sydney*'s was a midwinter cruise. Geographers, in describing the area, put it bluntly: 'Winters are cold. Indeed, frost is liable between October and April and in January the temperature rarely rises above freezing point, even in the daytime, though it has been known to reach 12.5°C.'[16] During the island campaign, temperatures of 9°F (-13°C) were recorded, and that was without the wind-chill factor. Aircrew would come below looking like Eskimos.

Although the flying crews found convoy escort boring compared with their normal attack roles, the pace of flying could be very taxing, operating from a relatively small deck in difficult midwinter conditions, against an enemy who could never be underrated. After two sorties a day for four consecutive days, the aircrews would be glad to see the replenishment ship coming over the horizon, as it would mean a day's rest and respite from flying. The varied tasks are illustrated by 2 January, the day that Sub-Lieutenant Coleman's Sea Fury was shot down. It was a day of poor visibility, though the land was blanketed in snow. Aircraft were flown off for 19 offensive sorties — two returned hit by flak. There were 14 combat air patrol and anti-submarine patrol sorties, 8 for target combat air patrol and 6 for search and rescue as a result of the non-return of Coleman from his CAP mission. Without their realising it, the strain was telling. CPO Frank O'Donovan recalled of Coleman's loss, 'We saw him fly into a cloud and disappear. The aircraft did not emerge from the other side.'[17] No-one has yet invented a Korean version of the Bermuda Triangle.

Nevertheless, there remains a laconic edge to recollections of the operations. Lieutenant Commander Lunberg narrates:

... in the afternoon Sub-Lt Dunlop, also in a Sea Fury iced up and was hit by flak on his way to Seoul. I was leading four Fireflies on 3 January 1952 when we went after some more bridges along the Chonsan–Chaeryong–Changyen line. Myself and Genge fixed the diversionary bridge while Bailey and Lee hit one to the north. The 5th saw one of the few successes at blocking rail tunnels. This one was north east of Ongjin and Lee put his bomb right in the entrance. Large puffs of smoke came out of the other end with suspected steam from the bombed end.[18]

By this time, frustration was setting in over the effect of Operation Strangle, and planes were sent off from the U.S. carriers in the small hours to take advantage of the moonlight on the snow. The clear air and the sparkling snow gave good visibility, at least on those few days when the weather was not too difficult. But still the supply lines were not strangled. As someone said, it was 'like trying to subdue an ant hill with a fly swat'.[19] For the *Sydney's* flyers, there was the knowledge that, for some time, in their area at least, all the bridges were destroyed, and the enemy was reduced to moving material by sheer manpower at night.

It was noted subsequently that air interdiction of enemy supplies had never proved a success in wartime, except in very isolated cases. It had been a notable failure in Italy in 1944, and Korea was to be no exception. Even then, the lesson was not learned, and the Americans attempted the same tactics — again a costly failure — in Vietnam, 15 years later.

Back in action on the seventh patrol, the airgroup was once more used against troops and railbridges. It was during this time that Sub-Lieutenant Rowland flew so low that he was hit by part of his own bomb as it exploded. Norman Lee was another whose aircraft was damaged by a piece of metal not fired by the enemy. The last sortie was on 24 January against troops near Yonan and Segori. Operations planned for the following day were cancelled because of blizzard conditions, force 9 winds, freezing spray, snow, and a temperature of -8 degrees Celsius. In these conditions, the working parts of the guns were liable to freeze solid and Aircraft were taken below decks to be warmed up for operation. It was an ironic that, prepared as she was for tropical service, the *Sydney* would go to war in effectively arctic conditions. (Korea is in an equivalent latitude to Bass Strait, but climatic conditions are vastly different.) On *Murchison*, too, visiting British officers noted the cold compared with their own ships, snugly insulated for possible use on Second World War convoys to Russia.

The *Sydney* left the war zone on 26 January 1952, to the great relief of all the aircrew, none of whom had looked forward to gaining the dubious distinction of being the last man shot down. She had been 118 days in

Korean waters, of which 89 were spent at sea, with 50 involving flying operations, including the escort of convoys to and from Inchon. Effectively, just over seven days' flying had been lost by bad weather.* She had steamed some 27 000 miles. The *Sydney*'s aircraft had made rocket, bombing and strafing attacks from the 38th parallel to the Yalu Gulf. Two thousand three hundred and sixty-six sorties were flown. Of these, 743 were by Fireflies, and 1623 by Sea Furies. These were about 72 per cent offensive, and the remainder defensive and reconnaissance. A similar allocation of sorties was experienced aboard the American aircraft carriers.

In summary, this is the record of the *Sydney*'s operations in Korea:

First Patrol	3–11 October	264 sorties
Second Patrol	17–27 October	387 sorties
Third Patrol	3–14 November	397 sorties
Fourth Patrol	18–30 November	280 sorties
Fifth Patrol	5–18 December	383 sorties
Sixth Patrol	27 December – 8 January	362 sorties
Seventh Patrol	15–26 January	293 sorties

Of her aircraft, ten had been shot down and 90 damaged by flak.†

One of result of *Sydney*'s tour of duty in Korea was the production of statistics in the American manner, the sort of thing ideal for cost-benefit analysis, not to mention the daily U.S. press briefing. Thus it was recorded that during her seven patrols, her airgroup had expended 269 249 rounds of 20 mm armament, 6359 rocket projectiles, and 902 bombs of 1000 or 500 pounds weight. The equivalent score for the *Glory* had been 573 000 rounds of 20 mm, 9240 rockets and 1514 bombs.

As for the *Tobruk*, she expended 2300 rounds of 4.5 inch and over 5000 40 mm shells. 'The quality of our marksmanship was very good,' said Commander Peek.[20]

The *Sydney* destroyed 66 bridges, 2 locomotives and 159 railway trucks, 2060 houses and 495 junks and sampans. The crew took it all in their stride, slightly bemused by the insistence of their ally on minutiae of doubtful statistical validity or practical use. Their humour extended to pointing out that the toll had also included *exactly* 234 ox carts destroyed and 155 damaged. It was not as trivial as it might have sounded. 'The Communists in Korea

---

* Some authors give different figures by including fractions of days lost.

† These figures differ in different sources. I have taken those from the newspaper reports immediately on her return to Australia. Given accidents and stress of weather, 13 aircraft were lost: 4 Fireflies and 9 Sea Furies.

use a lot of these narrow drays, each drawn by a single bullock. They look innocent enough, but many of them go off with a terrific bang when a few incendiary bullets hit them.'[21]

The *Glory*, on her tour, had destroyed 794! In addition to the ox carts, the peasants were dragooned into moving food and ammunition on big A-frames on their backs. Typically of all wars, the civilian population had suffered.

Three DSCs and one bar, and one DSM were awarded to her personnel, along with two U.S. Legions of Merit. *Sydney*'s casualties were listed as one killed, two missing and six wounded. The *Murchison*'s officers were also honoured, with two DSCs and a U.S. Legion of Merit. So honoured was the *Murchison* among her peers that, when she sailed for home, the crew of the *Bataan* lined the deck to cheer her out.

A count of structures destroyed or damaged is inadequate, not to say irrelevant, in measuring the *Sydney*'s success, but there was a job, and she did it. She was part of the big picture, when the free world stood up against militant Communism and drew the line. 'It has been a war to prevent a larger war by serving notice on a ruthless enemy that he can go so far and no farther.'[22]

And if the list of *Sydney*'s successes sounds less in relative terms than those of United States carriers, then this is probably due to the caution of Australian claims compared with the enthusiasm of the American. Unfortunately, no amount of air bombardment was going to change the course of the war. It required a decision on the ground.

One major result at least was her demonstrable efficiency, which earned the praise of senior commanders. At one time, Rear Admiral Scott-Moncrieff, and Vice-Admiral Harold Martin, commanding the U.S. Seventh Fleet, had both been aboard at the same time to watch the flying. They could attest that Australian ships stood up to the best traditions of any naval service in the world. There was also the sense of achievement and a job well done, as the only Australian aircraft carrier ever to go to war, that was to add to morale and tradition in later, leaner years.

The citation of the CBE awarded to Captain Harries later in the year reflected well on the whole crew:

> For devotion to duty while in command of HMAS Sydney, operating off the west coast of Korea for four months, during which time this most efficient carrier created a sortie record and consistently kept up a very high rate of sorties, which could only have been achieved by high efficiency of all hands from hard training under the supervision of Captain Harries. He displayed

excellent qualities of command and leadership under conditions of great strain and bad weather, all tasks asked for were accurately carried out.[23]

It is something of an irony that the bulk of training in all those elaborate manoeuvres in Australian and New Zealand waters was for situations that were never to occur in this real war. Interdiction of enemy supplies, and close support of the army — they were the proper use of naval air power in Korea.

The statement of the commander of one of the big U.S. carrier airgroups is also significant:

Generally speaking, the war in Korea demanded more competence, courage, and skill from the naval aviator than did World War II. The flying hours were longer, the days on the firing line more, the anti-aircraft hazards greater, the weather worse. There was less tangible evidence of results for a pilot to see.[24]

The Americans estimated that the anti-aircraft defence of certain targets was double anything that the Japanese had had during the Second World War. It was a major war, and the wear and tear on the American, as well as the allied forces, reflected it.

On 11 January 1952 the *Glory* and *Warramunga* sailed from Sydney for Korea, and six Fireflies were flown on board from the *Sydney* as they met at Hong Kong. Other aircraft and stores had been left with the *Unicorn*. From Hong Kong to Singapore, she transported a deck cargo of RAF Spitfires and Vampires.

One small curiosity of the Korean War was the operation of ships of two navies sharing the one name. The light aircraft carrier USS *Bataan* and the destroyer HMAS *Bataan* were both named in 1942, in honour of the heroic defence of the Bataan Peninsula, at the entrance to Manila Bay, which fell to the Japanese on 9 April 1942. The Australian ship was not completed in time to take part in the Second World War, but both were in Korea between December 1950 and June 1951, and again between August and September 1952. They operated in company from 13 to 25 March 1951.

Despite the use of Canadian destroyers in the Commonwealth force, there was great reluctance to commit the Canadian aircraft carrier. The *Sydney* and the British ships were configured as attack carriers, whereas the *Magnificent*, which had replaced her Fireflies with Avengers during 1950, was admitted to be useful only in an anti-submarine role. Such was the need, however, that a scheme was proposed whereby a squadron of twelve Canadian Sea Furies would be carried aboard a British ship, the *Warrior*, for service in Korea. With the ceasefire in July 1953, the matter went no further.

Absent from Korea except for a token force was France, which at this time was becoming more and more deeply enmired in its ruinous war in Vietnam. Unlike the Netherlands, which had protected its investments by renouncing sovereignty in Indonesia in 1949, the French fought doggedly to retain their colonies, and their great power status. The aircraft carriers *Arromanches* and *Lafayette* were stretched to their limit in support of the ground forces, with their Hellcat and Helldiver aircraft.

For the United States, the world leader in aircraft carrier technology, the Korean War was the opportunity the navy needed to prove that the carrier had a role in modern warfare. Smarting from the abrupt cancellation of the super-carrier *United States* in 1949, and bested in the contest to provide the launching platform for the nuclear deterrent, the navy was able to show its ships in a vital role of sustained non-nuclear strikes in support of ground forces. The war had saved the main carrier force from being reduced from eleven to eight units. The Eisenhower administration had been impressed. In 1952, despite objections from the Air Force, the attack carrier force was authorised at 14 units in place of its previous 12. The big aircraft carriers *Forrestal* and *Saratoga*, new ships laid down during the war, were the shape of things to come.

For Britain too, it was a time to boost the carrier force, though in her case there were wartime hulls that could be moved towards final completion.

Russia, although not officially involved in the war, was in the process of a major program of naval rearmament, which saw 13 cruisers and three battlecruisers of Stalin's rearmament program laid down in the three-year period. It had, at this time, only two cruisers at Vladivostok.

In Australia also, the Korean War was a boost for the navy.

9

THE WAY FORWARD

Things were busy in Sydney at the start of March 1952. On the 1st, 3000 people had cheered the launching by Mrs Menzies of the new destroyer *Voyager* at Cockatoo Island Dockyard. Royal approval had been given for the naming of the four new destroyers after the ships of the famous Scrap Iron Flotilla. Two days later 1250 soldiers of the 1st Battalion Royal Australian Regiment and 145 New Zealanders marched past the Governor-General, Sir William McKell, on their way to join the troopship *Devonshire* at Circular Quay, for Korea. They were sailing to join the 3rd Battalion in the Commonwealth Division under the United Nations Flag. Seven thousand people went to the Quay to see them off. Then, via Fremantle and Melbourne, the *Sydney* and *Tobruk* arrived home. The *Tobruk* was first, on the 4th, followed the next morning by the *Sydney* after she had flown off her aircraft to Nowra. Some 1500 people stood in the grey drizzle to greet her. The 6th saw two events, the first just a little poignant, as the liner *Gothic* berthed at Pyrmont with cargo following the end of the waterside workers' strike. She carried aboard all of the fittings that were to have been used for the Royal Tour, and the Royal Barge had to be unloaded from the top of one of the hatches. It had been intended that Princess Elizabeth should have landed in Sydney on 1 April, but for the death of the King. It was a good day, though, for the navy. Led by a company of the NSW Mounted Police, and accompanied by the RAN Band from Flinders Naval Depot, the ships' companies of the *Sydney*, *Tobruk* and *Murchison* marched through the streets of Sydney to mark their return from the war. The people gave them a 'ticker tape shower' as 900 men swung towards the Governor General on the steps of the Town Hall.[1]

The march was led by Commander Victor Smith of the *Sydney*, and at the Governor-General's request, the three commanding officers stood with him to take the salute. They were Captain David Harries of the *Sydney*, Commander Richard Peek of the *Tobruk*, and Lieutenant Commander Allen Dollard of the *Murchison*. Dollard's exploits in the Han River had been lauded by the press as having the Nelson touch. He also had the commendation of Admiral Scott-Moncrieff that 'No ship could have done better'.[2] Every man proudly wore, among his medal ribbons, two new ones, the yellow and blue of the British Commonwealth campaign medal, and the blue and white of the United Nations award.

The fourth naval member, Commodore A. R. Pedder, had congratulated the men in Melbourne, saying that the Australians had shown the world they could do it, and that he looked forward to a time when Royal Navy men would no longer have to help out aboard Australian warships. Captain Harries said, 'The men in this ship did a grand job', and the old hands, with their Second World War medals, said that 'the youngsters were splendid in all conditions'.[3]

Rear Admiral Scott-Moncrieff sent a final commendation, that 'Australia could be proud of *Sydney*, its first aircraft-carrier, and of a team that could not only achieve records but could maintain the highest possible standards in normal operating conditions in a war area'.[4]

Replacing the *Tobruk* and *Murchison* in the war zone were the *Bataan* and *Warramunga*. Both were to have eventful and stimulating patrols.

The stimulus of the Korean War produced a new program of Australian naval construction, announced in February 1952. It was to comprise:

6 fast anti-submarine frigates (River class)
1 fleet replenishment ship (*Tide Austral*)
3 boom defence vessels (Kimbla class)
4 coastal minesweepers (Bird class)
4 seaward defence boats (Ford class)
1 oil fuel lighter
2 deperming lighters
2 air/sea rescue vessels (Air class)★

In addition, it was decided that modernisation would proceed with the *Hobart* and the three Tribal class destroyers, as well as the conversion, already approved, of the Q class destroyers to fast anti-submarine frigates. Even 12 corvettes were to be brought out of reserve and upgraded. Work on the *Hobart* commenced in Newcastle in September 1952 after she arrived in tow of the *Wagga*. She was the biggest vessel to be refitted to date at the State Dockyard. Completion of the refit was to be at Garden Island, after which she was to replace the *Australia* in the fleet at the end of 1955. There was also a £539 000 program to modernise the facilities at Cockatoo Island Dockyard.

★ These last two were not in the official program but were built contemporaneously, one for the RAAF.

The Australian fleet was listed with the following strength for 1953:

Carrier force	2 light fleet carriers
	1 cruiser
	4 destroyers
Escort forces	1 "Q" class destroyer
	4 frigates
Surveying duties	2 survey ships and their tenders
Training ships	2 frigates
	8 fleet minesweepers
Auxiliary vessels	1 ocean-going tug
	1 ammunition carrier
	3 boom defence vessels
	3 air/sea rescue vessels
	2 patrol vessels
	2 general purpose vessels[5]

Personnel was to number 27 000, including 10 000 national service trainees, and citizen forces. In fact, the strength never reached anywhere near this high. In 1950 there were 10 252 permanent navy men, and 4943 citizen forces, and in 1952 the figures were 12 381 and 6254 respectively. 1953 saw the highest total until the 1960s, with 14 144 permanent and 7398 citizen forces. Included in these numbers were some 400 Royal Navy personnel on loan or exchange, while some Australians were in Britain. It was a source of some pride that Australian and British operational practices were almost identical.

Inevitably there was some carping in parliament that the defence budget was not big enough. William McMahon, the Minister for the Navy, however, stated:

> I consider that Australia has planned a well-balanced naval force which will be well capable of playing its part in conjunction with the forces of the United Nations, and particularly those of the United States of America and the United Kingdom. The government acts in close liaison with the United States Pacific fleet and the United Kingdom Pacific fleet and so far as the China Seas and Korean waters are concerned we have absolute superiority at the present time.[6]

Mr Riordan from the opposition benches pressed him to base patrol vessels at Cairns, Darwin and Manus, 'the three points of our northern triangle', to which Mr McMahon replied that another Fairmile motor launch was to be sent to the area. 'We in Australia are not without friends', he went on, and pointed out that there was no foreseeable threat such as the Japanese fleet had once posed, and that Australia had only to guard against submarines.[7]

The Australian plans were somewhat different from those of Canada, for instance, where developments were centred on the major escort role in the North Atlantic. This was the reason for the purely anti-submarine role of its aircraft carrier. The Australian fleet had a relatively unspecified role, except that it had to be as strong and as versatile as possible.

There was also the trans-Tasman connection. In December 1950, Vice-Admiral Collins had concluded fruitful discussions with New Zealand, and subsequently, the purchase of four Australian corvettes for the New Zealand Navy was raised. In the event, the ships were presented as a gift, and retained their Australian names, *Inverell*, *Echuca*, *Kiama* and *Stawell*. After exchanging the Australian flag for the New Zealand at Garden Island, they sailed for New Zealand between April and June 1952. The *Stawell* and two smaller minesweepers were used for Compulsory Military Training. In that year also, the *Bellona* sailed to England to participate in the big NATO exercise, Mainbrace. Except for its scale, involving fleets including two battleships and seven aircraft carriers among some 160 ships, this exercise was basically identical with those that had been carried out in Australian and New Zealand waters by the local forces.

The manpower strength of the New Zealand navy had been substantially increased since the late 1940s, and at the end of 1952 stood at 2662, including 102 on loan from the Royal Navy.

Four Australian GPVs and four motor launches were loaned to Britain for work on the Malayan coast in September 1951, and in June 1952 the first locally produced 21 inch torpedo was tested at Broken Bay.

For the aircraft carriers, steam catapults were ordered. William McMahon, pointed out that the ram and wire catapult had reached the limit of its developmental possibilities and that the new catapults would keep pace with aircraft development. Other new works set in hand were the building of new and improved accommodation at Manus Island, also HMAS *Watson*, Garden Island, (both in Sydney) HMAS *Harman* (Canberra) and HMAS *Lonsdale* (Melbourne).

The LSTs were all laid up by 1952, so the next year's Antarctic expedition used a chartered Norwegian sealer, the *Tottan*. The navy, however, was hoping for permission to build its own specialised Antarctic supply ship to support the planned permanent bases.

The provision of new fast escorts was becoming important, as the modern Soviet submarines of the Project 611 class had a submerged speed of 16 knots, a distinct advance on the modest 9 knots of the British T class boats in Sydney. British design studies by 1950 had evolved the type of fast anti-submarine frigate which was to be built for Australia. It required long-range, a good speed when driving into heavy seas, much improved electronics, and an armament envisaged as two 3 inch 70 calibre rapid fire AA guns, two 40 mm, 12 anti-submarine torpedo tubes and two triple Limbo ahead-throwing anti-submarine mortars. The Limbo was an advance on the Squid installed in the *Quadrant*. It had a range of 1000 yards and ammunition was carried for 20 salvoes. For all of these developments, the Australian industrial base was a major plus. Every new ship saw greater expertise in the dockyards, while the Government Ordnance factories at Maribyrnong and Bendigo were producing the most modern designs of gun and mounting. All of the *Anzac*'s guns had been manufactured here, and they do not seem to have suffered the teething troubles of the British guns aboard the *Tobruk*. Australia was well on its way to being able to stand on its own industrial feet.

When news was received of the death of the King on 7 February 1952, the *Australia* fired 56 minute guns, and then a 21-gun salute for the new Queen. In Korea, the funeral of the King was marked by a 56-gun salute, fired in an effective manner 'at the Queen's enemies' by the *Bataan* and *Cardigan Bay*.[8]

As part of her northern cruise that year the *Australia* carried Rear Admiral Eaton, Vice-Admiral Collins and Prime Minister Menzies from Brisbane to Townsville. *Anzac* was in company. Prime Ministers do not do such things nowadays, but Mr Menzies managed to produce a topical verse for each officer which he read out at his last dinner aboard the ship. The arrival of the *Australia* at Manus was a cause for celebration, as she brought new supplies of beer for the canteen. Then there was time to remember as she visited Guadalcanal, and tribute was paid to her lost sister ship, *Canberra*. Of course no-one was going to rest on their laurels, and the navy felt confident in challenging the air force at Williamtown to some competitive attacks on a piece of the local geography.

An important event of 1952 was the explosion of the first British atomic bomb, Operation Hurricane. Australia gave full and enthusiastic co-operation, and it was decided that the site would be the remote Monte Bello Islands, 72 kilometres off the north-west coast of Western Australia. Initial hydrographic work was performed, first by the *Karangi*, and then by the *Warrego*. Among the local features noted was the 18-foot tidal range. The *Warrego* was also surveying Exmouth Gulf in connection with oil exploration. Headquarters was to be the former escort carrier *Campania*, which had been seeing service as an exhibition ship for the Festival of Britain. The test was to be an exercise in the effects of a possible covert nuclear attack, by mine or torpedo, on a British harbour, and the nuclear device itself was to be located in a surplus frigate, the *Plym*. Two of the supporting landing craft sailed from Britain in February 1952, and the remaining ships in June, and all were in Fremantle at the end of July. Australia supplied some of the principal units of the navy to ensure that the whole area of the tests was clear and well patrolled. The prohibited area had a radius of 72 kilometres from Flag Island, and a further restricted area was proclaimed, to prevent observation of the tests by any unauthorised person. For the operation, *Sydney* had embarked one squadron of Sea Furies and two of Fireflies. These aircraft would be the key to the maintenance of the security of the area. Australian frigates would also assist with weather observation. Early in September the *Shoalhaven* and *Murchison* left Sydney with National Service trainees to join the *Macquarie* and *Thorough* in the Barrier Reef area for anti-submarine training. They then went on to join the *Sydney* and *Tobruk*, which had been operating out of Manus Island, as part of the force sent to Monte Bello.

Ships involved in the operation at Monte Bello were:

Australian fleet

Aircraft Carrier	*Sydney*
Destroyer	*Tobruk*
Frigates	*Shoalhaven*
	Murchison
	Culgoa
	Hawkesbury
	Macquarie
Landing ship	*Labuan*
Boom defence vessel	*Koala*
Tug	*Reserve*
Store ships	*MWL 251*
	MRL 252
Tenders	*Warreen*
	Limicola

British ships

Aircraft carrier	*Campania*
Frigate	*Plym*
Landing ships	*Narvik*
	Tracker
	Zeebrugge
Oiler	*Wave King*

The *Campania*, in her role as headquarters ship, carried two Sea Otters and three Dragonflies, for communication. The helicopters could take off from the deck, but the amphibians had to take off from the water. Much use was also made of the five LCMs and 12 LCAs from the British landing ships.

After some months of preparation, and waiting for appropriate weather, the *Plym* was vaporised on 3 October 1952. The men on the aircraft carrier had to face away from the explosion, and then were told when they could look at it safely. 'From "SYDNEY", we saw an orange flash the size of the rising sun light up the sky, followed by a ragged shaped atomic cloud, and then a terrific thundering noise caused the ship to vibrate.'[9]

For many, it was proof that Britain was still a great power. The *Campania* had stood off at a distance of 19 kilometres, but she and the other British ships were later found to have nuclear contamination in varying degrees. Although she had been intended to be converted for mercantile service on her return home, she spent some time in decontamination, and was then scrapped.

The Australian fleet, at a distance of 97 kilometres, was presumed not to have been affected.

Off Fremantle, on the return voyage, an unusual event took place. The weather was so rough that several Sea Furies which had flown overland from Nowra could be landed on the deck while the ship was lying at anchor. The ship had been lying to two anchors in a 40 knot wind, and as the aircraft approached one anchor was pulled up and she kept enough steerage way to keep the wind five degrees off the bow. For Tassie Webster and Jimmy Bowles it was the approach which was the most dangerous, as they had to avoid the masts of other ships lying at anchor in Gage Roads.

While the fleet was away, the air squadrons at Nowra were kept in form with exercises such as the sinking of the hulk of the former collier *Marjorie* off Sydney Heads. Nine Sea Furies and three Fireflies took 20 minutes to do that.

Subsequently, there was a rare piece of naval pageantry at Melbourne, as the ships came back from Monte Bello. HMAS *Culgoa* proudly flew three flags the blue ensign of Australia, the white ensign of the Queen's navy, and the special flag of the Australian Naval Board for the occasion, as she received the salutes of the fleet at 10 a.m. on 30 October. She lay off Black Rock, not far from the hulk of one of the progenitors of the navy, the old Victorian monitor *Cerberus*. The sailors went, as reporters put it, 'breezily to breakfast',[10] and the fleet, which had been at anchor off Mount Martha, got under way at 8 a.m. Then, with their crews smartly lining the decks, they came up

Port Phillip in line ahead at 8 knots. The flagship *Sydney* was in the lead, her band playing 'Rule Britannia', and 16 of her aircraft roaring overhead in an anchor formation. Rear Admiral Eaton returned the salutes of Vice-Admiral Sir John Collins and Commodore Dowling with 15 guns. Behind in order came the *Australia, Tobruk, Shoalhaven, Murchison, Macquarie, Latrobe, Wagga, Cootamundra, Gladstone, Colac* and *Cowra*, and under a grey Melbourne sky they made a fine showing. Cars on Beach Road were crowded thicker than a summer holiday as the people came out to see. Then, when the ships were alongside the piers, the sailors headed off to the long-considered delights of the city and St Kilda. Ten days alongside was time for a decent blowout. The *Culgoa* for her part went back to sea for a gunnery shoot.

Aboard the *Australia* that night, Admiral Collins entertained 500 guests while many more, less official, climbed all over the ships on the weekend. There was also Operation Flemington, for which Morse Code was again the favourite in the Melbourne Cup. Once again it was not the winner. (The lucky ticket had Dalray written on it.) Melbourne on Cup day had a real holiday atmosphere. There were 90 000 people at Flemington, and another 30 000 visited the *Sydney, Australia* and *Tobruk* at the Port Melbourne piers.

The new ships came forward slowly. Two Daring class destroyers were laid down in 1949, and a third in 1952. They featured the first all-welded warship hulls to be built in Australia, the machinery required new standards of dimensional accuracy, and they also showed greater sophistication in electronics and weapons control. In 1950, the conversion of the Q class destroyers to fast anti-submarine frigates commenced. The first, *Quadrant*, recommissioned in July 1953. Other ships were modified for an improved anti-submarine capability. The significant delay involved the second aircraft carrier, the *Melbourne*, and to bridge the gap, the Royal Navy offered the loan of HMS *Vengeance* until she should be ready. A new generation of naval aircraft had been ordered, the Sea Venom in 1951 and the Gannet in 1952. After a two-month dockyard refit, the *Vengeance* commissioned into the RAN on 13 November 1952, left the Clyde with a steaming party of 550 men aboard on 22 January 1953, and berthed at Princes Pier, Melbourne, on 5 March. The old *Australia* was at Station Pier to greet her. She arrived in Sydney on 11 March. Here the rest of her crew and the flying officers, joined her. *Vengeance* was insulated for tropical service and partly air-conditioned. Her previous mixed anti-aircraft armament was now a standard sixteen 40 mm guns. She also brought with her the navy's first three Bristol Sycamore helicopters, a great improvement on the old Sea Otter amphibians.

The *Melbourne* was to be the recipient of the three major advances in aircraft carrier technology of the early 1950s, the angled deck, mirror landing gear, and the steam catapult. These were the breakthroughs that made possible the wholesale use on carriers of high-performance jet aircraft. All three were British innovations, though the United States was to be the main beneficiary. First experiments in angled decks took place during 1952, and the first mirror landing gear was tried towards the end of the year. By 1953 it was evident that this was the way of the future, and the *Melbourne* was to be the first ship of her type to have the benefit of all the modern advances.

The air base at Schofields was transferred from the RAAF to the RAN on 1 April 1953 as HMAS *Nirimba*. The air force then transferred their operations to the base at Richmond, which could be developed better for jet fighters and high-speed aircraft. The naval requirements, it was stated, were less exacting. The base, usefully located in the outer suburbs of Sydney, was to house two naval reserve squadrons, a repair base, and the Captain (Air) Australia.

The arrival of a second operational aircraft carrier meant that naval policy could achieve a greater flexibility. One big ship, for instance, would be available for local operations while the other could be deployed overseas. The *Vengeance* took her aircraft to sea for the first time in June 1953. 805 and 850 Squadrons of Sea Furies and 816 of Fireflies then began working up with the intention that the *Vengeance* should go to Korea. 850 Squadron had been commissioned at Nowra on 12 January 1953.

There had also been a quiet change in Australian defence policy. As France's position in Vietnam deteriorated during 1952, Prime Minister Menzies backed away from any commitment of Australian forces to the Middle East in a major war. They would be needed much more urgently in our part of the world. Indeed, by late 1953 it had been agreed that all Australian forces would be available for service in Malaya, where they would also have the support of the New Zealand navy and air force. It was the concept which was to give birth to the proposal for the creation of a 'Far East Strategic Reserve'.

The *Sydney* was Australia's representative at the Coronation Fleet Review on 15 June 1953. She sailed from Melbourne for the event, under the command of Captain Herbert Buchanan, escorted by HMNZS *Black Prince*, under Captain G. Dolphin, RN, on 24 March. Only one squadron was embarked, No. 817 of eight Fireflies, under Lieutenant Commander Albert Oakley. It was assured that the ship would be 'proper tiddley' for the event.[11] The band played and Admiral Collins took the salute from the end of Station Pier as the ships steamed out, bound for Fremantle, Colombo, Aden and the Suez Canal.

On the way, there were exercises with the Mediterranean Fleet, when *Sydney*'s Fireflies mixed with the Fireflies and Sea Furies from Admiral Mountbatten's flagship *Indomitable* in attack and defence. They then went on to a joint attack on Gibraltar on 30 April, and the 'ANZAC Squadron' arrived in Portsmouth on 5 May.

Fleet reviews have more than a ceremonial function. In this case, it was time for the navy to reassert its role in the defence of the realm, a role that had been severely discounted by the glorification of such things as the role of the Royal Air Force in the Battle of Britain. The previous review had been for the coronation of George VI in 1937, and the intervening 16 years had seen many changes, notably the rise to prominence of naval aviation. Nevertheless, most commentators were fully aware of what they were seeing, and noted, 'It is unfortunately only too well known that the Fleet Air Arm is passing through a weak phase' as well as how sadly reduced was the size of the fleet.[12]

Still, it was a great show, displaying 'the wonderful efficiency of the officers and men of the Royal Navy'.[13] Perhaps it was the last of the really great reviews, and the names of the principal warships read like a litany of the prestige of what had once been the world's greatest navy: *Vanguard, Eagle, Implacable, Indefatigable, Illustrious, Indomitable, Sydney* (RAN), *Magnificent* (RCN), *Theseus, Perseus, Devonshire, Glasgow, Sheffield, Superb, Ontario* (RCN), *Gambia, Swiftsure, Quebec* (RCN), *Delhi* (India), *Dido, Cleopatra, Black Prince* (RNZN), *Apollo, Manxman, Decoy, Defender, Duchess* and *Diamond*. Through the lines of ships, 229 in all, of which 200 were from the Royal Navy, the young Queen sailed to take their salute. Vice-Admiral John Eccles, Flag Officer Air (Home), was present aboard the *Illustrious*, and Admiral Sir George Creasy, Commander in Chief of the Home Fleet, was on the *Vanguard*. Significant among the foreign ships present were the *Sverdlov* (USSR), *Baltimore* (U.S.A.), *Almirante Barroso* (Brazil), *Miguel de Cervantes* (Spain), *Montcalm* (France), *Gota Lejon* (Sweden) and *Tromp* (Netherlands). Observers of a historical bent noted that it was just two months short of a hundred years since Queen Victoria had come to this same place to review 50 ships of the Channel Fleet, including ten battleships and 14 frigates, mostly steamers, their sides bristling with no less than 1151 guns. Now it was aircraft that were important, with a flypast in the late afternoon of over 300 planes from 37 squadrons, including five squadrons of jets and one of helicopters. Victoria was 34 years of age as she stood on the deck of the yacht *Victoria and Albert* to 'reckon up her battleships'. The new Queen aboard the despatch vessel HMS *Surprise* to acknowledge the salutes of her loyal sailors was aged 27.

The return of the *Sydney* to Australia in August was via Halifax, Annapolis, Panama, Pearl Harbor and Auckland. During the Atlantic crossing she participated in exercises with five Canadian warships, the whole force under Rear Admiral R. E. S. Bidwell, commanding the Canadian Coronation Squadron. There was wild weather, with waves coming over the flight deck and a possibility that the ships would have to return to port. On another occasion, due to persistent fog, both the *Magnificent* and *Sydney* had difficulty retrieving their aircraft. In Kingston, Jamaica, the local water was suspect, so the *Sydney* made use of its own desalination plant. The sailors noted the muddy condition of the end product and suspected that it was sourced from somewhere very close to the bottom of the harbour. They also watched carefully as the ship was squeezed through the locks of the Panama Canal with just inches to spare. For her part, the *Black Prince* returned to Auckland in September via exercises with the Mediterranean Fleet.

The *Vengeance* meanwhile operated her squadrons in Australian waters. 805 and 816 were embarked during the winter of 1953, followed by 808 and 817 after the return of the *Sydney* in the spring and summer. In August 1953 she acted as guide for aircraft participating in the London to Christchurch air race.

It is time for a review of the Royal Australian Navy here in 1953, at one of the high points of its existence. Building and refitting of ships for the conditions of modern warfare was well in hand, and the sight of our two aircraft carriers at Garden Island was an indication that the main goals of the 1947 plan had been achieved. The fleet would be well-versed in the tactics of anti-submarine warfare by now, with exercises such as that which followed the visit to the Hobart Regatta. The whole event was coordinated by the Australian joint Anti-Submarine School at Jervis Bay. Ships taking part were:

Australian fleet
Aircraft carrier	*Sydney*
Cruiser	*Australia*
Destroyers	*Tobruk*
	Arunta
	Bataan
Frigates	*Shoalhaven*
	Murchison
	Macquarie

British ships
Submarines	*Tactician*
	Thorough

The destroyers and the *Macquarie* were able to practice with their Squid anti-submarine mortars. Neither the submarines nor the frigates were the sharp end of technology, but with the recommissioning of the *Quadrant* in July as senior ship of the first Australian frigate squadron, this was about to change. There would be other duties around the coast, with the *Barcoo* replacing the *Warrego* on the Western Australian surveys, and the resumption of the northern frigate patrol against illegal Japanese fishing. The Japanese fisheries were expanding hugely and the tonnage caught was to double in the 20 years after 1948. For the protection of our resources, surveillance was necessary. The *Macquarie* was replaced on station by the *Shoalhaven*, and then by the *Hawkesbury*, which also found time to tour the islands to visit the coastwatchers. Manus was a poor base for this work that required patrols to concentrate on the waters around Bathurst Island and the north-west coast of Western Australia. Although four frigates would ensure a permanent presence, a base at Darwin would have been preferable.

Regular deployment of ships to patrol the ceasefire in Korea continued. The *Culgoa* relieved the *Condamine*, and was in turn relieved by the *Murchison*. The *Tobruk* relieved the *Anzac*. There was no shooting this time, just vigilance and the display of power. It was the *Sydney*, however, rather than the *Vengeance*, which was to return towards the end of 1953 to relieve the *Ocean*.

New Zealand also kept up a modest force, with ships passing into reserve and being replaced by others for operational purposes. The *Black Prince* remained with the Royal Navy in the Mediterranean until September 1953 specifically because of the operational experience to be gained, and at the same time, the duties of the frigates sent to Korea were widened to include the whole of the Far East Station.

As for the future, it was to be affected by the highest inflation recorded in Australia in modern times, resulting in significant moves in government policy, retrenchment, and other problems. For the navy, the 1954 Defence Review included substantial changes. And as the cancellations and the slowing of projects took effect, it was plain that the future of the navy was never again to be as confident as it had been in those immediate postwar years. There was just that moment of achievement when the Fifth Aircraft Carrier Squadron, *Sydney* and *Vengeance*, operated their aircraft together for the first time in Hervey Bay in September 1953. In company were the *Anzac*, *Arunta* and *Quadrant*. The *Sydney*, under Captain George Oldham, was working up to go to Korea, and the *Vengeance*, under Captain Henry Burrell, was there to assist. Hervey Bay was a place where the weather could be relied upon to provide good flying conditions. As a result, when on their steaming south in October, the old boom gate vessel *Kuramia* was towed out as a target ship by the *Wagga*, it was sunk by the third aircraft. The *Sydney* and *Bataan* would fly Australia's flag again in the wintry northern waters, the *Vengeance*, would prepare to accompany the Queen on her first visit to this side of the world, and the *Australia* herself would prepare, if not to carry the monarch, at least to provide proud escort to the Royal Yacht *Gothic*.

For the process started in 1947, it was a culmination.

It is fitting to end with a song.

> 805 flew in from Nowra
> To embark on a tropical cruise
> We were greeted in *Vengeance* the right way
> The Fish-heads bought plenty of booze
> Oh land us on, Hughie, land us on.[14]

AFTERWORD

The future is rarely predictable. In 1945 Douglas MacArthur broadcast in grand phrases.

> To the Pacific basin has come the vista of a new emancipated world. Today, freedom is on the offensive, democracy is on the march. Today in Asia as well as in Europe, unshackled peoples are tasting the full sweetness of liberty.[1]

Freedom, when it came, was often the result of further armed struggle, and sometimes had the name of communism attached to it. The result was the creation of further defensive alliances, one of which was the 1954 South East Asia Collective Defence Treaty (known as SEATO), which turned Australian eyes further towards involvement in Asia. From the American point of view, it was an anti-communist treaty pure and simple, and it promised, as well as the defence of its signatories, to include Cambodia, Laos and free Vietnam. For Australia it promised both the security of further alliances and the possibility of having once again to fight in a war in a foreign country. The *Vengeance*, *Anzac* and *Bataan* had visited ports in the region after concluding their escort of the Royal Yacht. There was also what might be described as the gentlest of implied hints in the visit to Australia every year from 1954 of significant American naval units for the celebrations of the anniversary of the Battle of the Coral Sea. The visit in that year of the aircraft carrier *Tarawa* to Sydney and Melbourne also saw speeches by Fleet Admiral William Halsey. Admiral Collins said that the Coral Sea Battle, 12 years before, was as important to Australia as Trafalgar to the British, and Halsey pointed to an old truth, that 'weakness invites warlike strength and preparedness forestalls it'.[2]

The regular stationing of ships in Singapore was part of what was called the Commonwealth Strategic Reserve commenced in mid-1955. *Arunta* and *Warramunga* were our first contribution. In practice it provided a stronger force on the Royal Navy's Far East Station than the Royal Navy itself could supply, and it was an assurance of the presence of adequate force in a part of the world that seemed to be subject to continual turbulence. The *Melbourne*, after she finally arrived in Australia on 23 April 1956, was to participate in regular exercises with ships on this station right from the outset. Indeed, because of the requirements of the Royal Navy in fighting in the Mediterranean, the Australian force which visited Singapore, Hong Kong and Manila in September and October 1956 for Exercise Albatross was far the strongest part of the Commonwealth contingent. It comprised *Melbourne, Sydney, Anzac, Tobruk, Quadrant, Queenborough* and *Quickmatch*.

Sometimes it almost came to shooting. On one occasion the *Anzac* was seen as providing de facto convoy escort for the Dutch passenger liner *Nieuw Holland* through Indonesian waters.

Naturally, the focus on Singapore meant that Manus Island was able to slip away after 1956 into a permanent somnolence. Not only was it strategically unsound, it was irrelevant. The New Zealand link, also, moved more into the framework of the wider defence context.

Conscription for the navy finished in 1956, and patrol work in Korean waters became just one of the functions of one of the ships on the Commonwealth Strategic Reserve, a function soon to be abandoned. Naval manpower levels began to sink. Another change was heralded by the first visit of the *Melbourne* to Pearl Harbor in 1958. The crew on the flight deck lined up to spell out the word 'Aloha'. After so long firmly confined within the British Commonwealth defence orbit, Australian eyes were looking towards our other great and powerful friend.

So Australia slipped back into the comfortable and traditional role of a dependable ally, with forces committed piecemeal as required. The fleet became more and more concentrated on its anti-submarine escort role, even to the extent of a Defence Review in 1959 proposing to phase out the whole naval air arm within four years. Fortunately, that decision was reversed, and the carrier *Melbourne* was to be the inheritor of a great tradition, and the flagship of the Australian fleet, for more than another two decades.

Today's navy can still look back with pride.

Glossary

Avgas Aircraft petrol

Bofors A light automatic rapid-firing gun of Swedish design, in this case, with a calibre of 40 mm.

CPO Chief Petty Officer

Cutter Naval pulling (rowing) boat, 32 feet long

DSO Distinguished Service Order. Service slang would have it as an award for the wartime loss of a vital part of the anatomy. DSC and DSM are different grades of the award.

Fairmile A type of naval motor launch evolved during the Second World War for use as a gunboat or torpedo boat. The usage by Mr McMahon on page 96 was incorrect, as all Fairmiles had been sold out of the RAN at the end of the war.

Flak anti-aircraft fire

Frigate Escort vessel evolved during the Second World War. The sloop and corvette are similar types of vessel.

GPV General Purpose Vessel. A small wooden naval cargo vessel.

Grand National A steeplechase horse race. To crash and entangle the wheels of an aircraft in the crash barrier.

Greenie Members of the electrical branch were so called, not because of their conservationist tendencies, but because of the colour of their insignia.

Gunroom Very junior officers' mess

HDML Harbour Defence Motor Launch

Hedgehog An ahead-throwing anti-submarine weapon using contact bombs

HMAS His Majesty's Australian Ship (Her Majesty's from 6 Feb. 1952). Thus, HMS, HMCS and HMNZS for British, Canadian and New Zealand warships.

Junk Sailing vessel of Chinese design.

Knot Unit of speed. 1 nautical mile (1.853 kilometres) per hour.

LCM Landing Craft (Mechanised). The LCA (Landing Craft Assault) was basically similar, both being designed to carry a platoon of soldiers or a single tank.

LST Landing Ship (Tank)

Matelot French term for a sailor, frequently used in English.

Mess Eating area on a ship, and by extension, members of a group who eat together.

MWL Motor Water Lighter. A small steel naval store carrier, in this case for fresh water. The MRL was an identical vessel equipped with refrigerated storage.

NATO North Atlantic Treaty Organisation, established in August 1949.

NAVAL BOARD FLAG Divided horizontally with blue below and red above, with a yellow foul-anchor superimposed horizontally.

OBE Order of the British Empire. CBE is a different grade of the order.

PARAVANE A kind of water kite which, when towed, will divert cables and other gear away from a ship's side.

PO Petty Officer

PONGO Naval slang for a member of the army.

PUSSER'S Belonging to the navy

RAAF Royal Australian Air Force. Thus also RAF and RNZAF for Britain and New Zealand.

RAN Royal Australian Navy. Thus also, RN, RCN and RNZN for the navies of Britain, Canada and New Zealand.

RANR Royal Australian Naval Reserve

RANVR Royal Australian Naval Volunteer Reserve

RAR Royal Australian Regiment

RFA Royal Fleet Auxiliary. A merchant-manned support ship.

RNAS Royal Naval Air Station

SALVO Simultaneous discharge of weapons

SAMPAN Small vessel of Chinese design

SPONSON A projection from a ship's side, eg for a gun platform.

SQUID An ahead-throwing anti-submarine weapon using depth charges. Limbo was a more advanced version.

STOKER No longer 'the joker who works with a shovel and a poker', the term applied to sailors working the engines of the ship.

TONNAGE All data is in English tons displacement (1 ton = 1.02 tonnes)

USS United States Ship. Thus also USN for United States Navy.

WARDROOM Officers' mess

WHALER Naval pulling (rowing) boat, 27 feet long, pointed at both ends

WRANS Women's Royal Australian Naval Service. Thus also WRNS and WRNZNS for comparable British and New Zealand services.

Particulars of Principal Warships

Displacement (full load), length × breadth × draught, maximum speed
Armament (number and calibre of guns, torpedo tubes, number of aircraft and other weapons)

BATTLESHIP
New Jersey
57 540 tons   887'3" × 108'2" × 38'   32.5 knots
9-16" 20-5" 76-40 mm 2 helicopters

AIRCRAFT CARRIERS
Valley Forge
36 380 tons   862' × 108' (deck) × 27'   33 knots
87 aircraft 10-5" 44-40 mm

Melbourne
20 000 tons   701'6" × 126' (deck) × 25' 24 knots
22 aircraft 25-40 mm

Sydney and *Magnificent*
19 550 tons   690' × 80' (deck) × 24'   24.5 knots
37 aircraft 30-40 mm

Glory, Ocean, Theseus, Vengeance and *Warrior*
18 040 tons   690' × 80' (deck) × 23'   25 knots
39 aircraft 24-2pdr 18-40 mm

CRUISERS
Belfast
14 930 tons   613'6" × 66'4" × 23'   32.5 knots
12-6" 8-4" 32-2pdr 9-40 mm 6-21"TT

Australia
13 630 tons   630' × 68' × 22'   31 knots
6-8" 8-4" 12-40 mm

Hobart
9 420 tons   562' × 56'8" × 20'   32.5 knots
6-6" 8-4" 8-2pdr 9-40 mm 8-21"TT

Bellona and *Black Prince*
7 410 tons   512' × 51' × 18'   30 knots
8-5.25" 6-40 mm 6-21"TT

DESTROYERS
Anzac and *Tobruk*
3 375 tons   379' × 41' × 13'6"   32 knots
4-4.5" 12-40 mm 10-21"TT 1 Squid

Arunta, Bataan and *Warramunga*
2 675 tons   377'6" × 35'8" × 15'6"   36 knots
6-4.7" 2-4" 6-40 mm 4-21"TT

Cockade and *Contest*
2 525 tons   362'9" × 35'8" × 14'   33 knots
4-4.5" 4-40 mm 4-21"TT

FRIGATES
Quadrant
2 700 tons   358'9" × 35'9" × 15'6"   32 knots
2-4" 2-40 mm 2 Squid

Hawea, Kaniere, Pukaki, Rotoiti, Taupo and *Tutira*
2 260 tons   307'6" × 35'8" × 14'3"   19 knots
1-4" 4-2pdr 4-40 mm 2 Squid

Condamine, Culgoa, Murchison and *Shoalhaven*
2 106 tons   301'8" × 36'6" × 13'   19.5 knots
4-4" 3-40 mm 1 Hedgehog

Barcoo, Lachlan and *Macquarie*
2 074 tons   301'8" × 36'6" × 13'   20 knots
2-4" 3-40 mm

SUBMARINE
Tactician, Thorough and *Telemachus*
1 090/1 571 tons 273'6" × 26'6" × 15'   15/9 knots
1-4" 11-21"TT

SUPPORT CARRIER
Unicorn
20 300 tons   646' × 90' × 25'   24 knots
8-4" 16-2pdr 16-20 mm, capacity for 20 aircraft

TROOPSHIP
Kanimbla
5 618 tons   494' × 66'3" × 24'3"   19 knots
capacity for 1280 troops

LANDING SHIP
Labuan
3 117 tons   345'10" × 55'3" × 11'6"   13 knots
capacity for 18 tanks and other equipment

Particulars of Principal Aircraft

Hawker Sea Fury FB11
Engine — 1 × 2550 hp Centaurus
Loaded weight — 12 500 lbs
Speed — 460 mph @ 18 000 feet
Ceiling — 35 800 feet
Climb — 10.8 min to 30 000 feet
Armament — 4-20 mm cannon 2 - 1000 lb bombs or 12 rockets

CAC Mustang CA18
Engine — 1 × 1400 hp Merlin
Loaded weight — 11 600 lbs
Speed — 437 mph @ 14 000 feet
Ceiling — 41 900 feet
Climb - 13 min to 30 000 feet
Armament - 6-.5" MG 2-1000 lb bombs or rockets

Fairey Firefly FR5
Engine - 1 × 2250 hp Griffon
Loaded weight - 16 096 lbs
Speed - 386 mph @ 14 000 feet
Ceiling - 28 400 feet
Climb - 15.5 min to 20 000 feet
Armament - 4-20 mm cannon 2-1000 lb bombs or 16 rockets

Supermarine Seafire XV
Engine - 1 × 1850 hp Griffon
Loaded weight - 8000 lbs
Speed - 383 mph @ 13 500 feet
Ceiling - 35 500 feet
Climb - 7 min to 20 000 feet
Armament - 2-20 mm cannon 4-.303" MG 1-500 lb bomb or 8 rockets

Fairey Firefly FR1
Engine - 1 × 1990 hp Griffon
Loaded weight - 14 020 lbs
Speed - 316 mph @ 14 000 feet
Ceiling - 28 000 feet
Climb - 5.8 min to 10 000 feet
Armament - 4-20 mm cannon 2-1000 lb bombs or 8 rockets

Supermarine Sea Otter
Engine - 1 × 855 hp Mercury
Loaded weight - 10 830 lbs
Speed - 150 mph @ 5000 feet
Ceiling - 16 000 feet
Climb - 6.2 min to 5000 feet

Korean War and other service by Australian and New Zealand vessels on the Far East Station

AUSTRALIAN	
Hobart	4/47 – 7/47
Quiberon	4/47 – 7/47
Quickmatch	4/47 – 8/47
Kanimbla	5/47 – 6/47
Culgoa	6/47 – 12/47
Kanimbla	7/47 – 9/47
Bataan	8/47 – 11/47
Australia	9/47 – 11/47
Manoora	9/47 – 11/47
Arunta	11/47 – 3/48
Warramunga	11/47 – 3/48
Kanimbla	2/48 – 3/48
Quickmatch	3/48 – 7/48
Quiberon	3/48 – 7/48
Bataan	7/48 – 11/48
Warramunga	10/48 – 2/49
Kanimbla	12/48 – 1/49
Shoalhaven	2/49 – 6/49
Bataan	5/49 – 9/49
Culgoa	8/49 – 2/50
Shoalhaven	1/50 – 9/50
Bataan	6/50 – 6/51
Warramunga	8/50 – 8/51
Murchison	5/51 – 2/52
Anzac	8/51 – 10/51
Sydney	9/51 – 3/52
Tobruk	9/51 – 3/52
Warramunga	1/52 – 7/53
Bataan	1/52 – 9/52
Condamine	7/52 – 3/53
Anzac	9/52 – 6/53
Culgoa	3/53 – 11/53
Tobruk	6/53 – 2/54
Sydney	11/53 – 5/54
Bataan	11/53 – 1/54
Murchison	12/53 – 7/54
Arunta	1/54 – 10/54
Shoalhaven	7/54 – 3/55
Vengeance	10/54 – 11/54
Anzac	10/54 – 4/55
Condamine	3/55 – 10/55
Arunta	6/55 – 12/55
Warramunga	6/55 – 12/55

NEW ZEALAND	
Pukaki	7/50 – 11/50
Tutira	7/50 – 5/51
Rotoiti	10/50 – 11/51
Hawea	3/51 – 2/52
Taupo	9/51 – 10/52
Rotoiti	1/52 – 3/53
Hawea	8/52 – 8/53
Kaniere	3/53 – 2/54
Pukaki	10/53 – 9/54
Kaniere	10/54 – 9/55
Black Prince	5/55 – 7/55

PRICES AND INFLATION

Inflation jumped from 6% in 1948 and 9% in 1949 and 1950 to 13% in 1951 and 23% in 1952. It was combated by government fiscal policies which brought on a severe recession with zero economic growth in 1952, and a contraction of 3% in 1953.

£3 million in 1947-48 would be equivalent to $120 million in 1995-96.

THE SOVIET SUBMARINE FORCE

Postwar thinking on submarines was dominated by memories of the German wartime campaigns, and also the phenomenal wartime success of U.S. submarines against Japan. It was imagined that the large Soviet submarine force would be unleashed to cripple Western shipping and communications. But, despite the numbers of boats involved, the Soviets, at least in the 1950s, would have provided a very poor threat. In the Great Patriotic War, Russian submarine commanders, with poor training and equipment, but faced with the necessity of success, claimed the sinking of almost four times the number of ships than was actually the case. In strategic terms, they were seen as an extension of the Red Army, and their record has been characterised as 'undistinguished'.[*] The laying of defensive minefields was a major activity. In 1949 there was a purge of submarine officers in the Northern Fleet.

Always arranged for defence, and strongest nearer their bases, the Soviet Pacific Fleet in the Korean War period could rely on only about 17 of its total of 84 boats for extended cruising, located at six bases. Perhaps half of these may have been manned and capable of operation.

Things would change, but the Soviet submarine strength would always be a substantial bluff.

★ Rolf Erikson, 'Soviet Submarine Operations', in J. Sadkovitch, (ed.), *Reevaluating Major Naval Combatants*, p. 174

BIBLIOGRAPHY

NEWSPAPERS
The Illustrated London News
The Times
The Navy
The Sydney Morning Herald
The Age
The Sun News-Pictorial
The Herald
The Argus
Port of Melbourne Quarterly
The Courier Mail
New Zealand Herald
The Dominion
The Press
The Mercury

GOVERNMENT PUBLICATIONS
Australia in Facts and Figures
Commonwealth of Australia, Parliamentary Debates
New Zealand Official Year-Book
Pilot's Notes for Sea Fury Mks. 10 and 11, Admiralty, May 1950
Yearbook of the Commonwealth of Australia

BOOKS AND ARTICLES
Anderson, Peter N, *Mustangs of the RAAF and RNZAF*, Reed, Sydney, 1975.
Barclay, Glen St J., *Friends in High Places*, Oxford University Press, Melbourne, 1985.
Bartlett, Norman, *With the Australians in Korea*, Australian War Memorial, Canberra, 1954.
Bastock, John, *Australia's Ships of War*, Angus & Robertson, Sydney, 1975.
Beaver, Paul, *The British Aircraft Carrier*, Patrick Stephens, Wellingborough, 1984.
Bell, Coral, *Dependent Ally*, Oxford University Press, Melbourne, 1988.
Bird, Peter, *Operation Hurricane*, Square One, Worcester, 1989.
Boughton, T.W. & Parnell, N. M., 'RAN Fleet Air Arm Aircraft', in *Aviation Historical Society of Australia Journal*, Vol. XV, No.1, March–April 1974.
Bovey, John, 'The Destroyers' War in Korea, 1952-53', in James Boutelier, (ed.), *RCN in Retrospect*, UBC, Vancouver, 1982.
Brown, Eric, *Wings of the Navy*, Airlife, Shrewsbury, 1987.
Brown, J. D., *Carrier Operations in World War II, Vol. 1*, Ian Allan, London, 1968.
Burgess, M., *Aircraft Carriers and Aircraft-Carrying Cruisers*, Burgess Media, Dunedin, 1980.
Cable, James, *Navies in Violent Peace*, Macmillan, Basingstoke, 1989.
Cable, James, *Gunboat Diplomacy 1919-1991*, Macmillan, London, 1994.
Cagle, Malcolm and Manson, Frank, *The Sea War in Korea*, USNI, Annapolis, 1957.
Cassells, Vic, *For Those in peril...*, Kangaroo Press, Sydney, 1995.
Chesneau, Robert (ed.), *Conway's All the World's Fighting Ships 1922-46*, Conway, London, 1980.
Collins, Sir John, *As Luck Would Have It*, Angus & Robertson, Sydney, 1965.
Cooney, David, *A Chronology of the U.S. Navy 1775-1965*, Watts, New York, 1965.
Cornish, W. B., *Monsoon Lands, Part 2*, University Tutorial Press, London, 1967.
Coronation Review of the Fleet, Official Souvenir Programme, Gale & Polden, Portsmouth, 1953.
Crosley, R. M., *Up in Harm's Way*, Airlife, Shrewsbury, 1995.
Dictionary of American Naval Fighting Ships, U.S. Government Printing Office, Washington, 8 vols, 1959-81. (Aircraft are covered in an appendix in Vol. 5.)

BIBLIOGRAPHY

Donohue, Hector, *From Empire Defence to the Long Haul*, Department of Defence (Navy), Canberra, 1996.

Edwards, Peter, *Crises and Commitments*, Allen & Unwin, Sydney, 1992.

Erikson, Rolf, 'Soviet Submarine Operations' in James Sadkovich, (ed.), *Re-evaluating Major Naval Combatants of World War II*, Greenwood, New York, 1990.

Frame, Tom, *Pacific Partners*, Hodder & Stoughton, Sydney, 1992.

Friedman, Norman, *U.S. Aircraft Carriers*, Arms & Armour, London, 1983.

Friedman, Norman, *The Postwar Naval Revolution*, USNI, Annapolis, 1986.

Friedman, Norman, *British Carrier Aviation*, Conway, London, 1988.

Garbutt, Paul, *Naval Challenge 1945-1961*, Macdonald, London, 1961.

Gardiner, Robert (ed.), *Conway's All the World's Fighting Ships 1947-95*, Conway, London, 1995.

German, Tony, *The Sea is at Our Gates*, M&S, Toronto, 1990.

Gillett, Ross, *Australian and New Zealand Warships since 1946*, Child, Sydney, 1988.

Gillett, Ross, *Wings Across the Sea*, Aerospace, Sydney, 1988.

Goldrick, James, 'Carriers for the Commonwealth' in T. Frame, J. Goldrick and P. Jones, (eds), *Reflections on the RAN*, Kangaroo Press, Sydney, 1991.

Goldrick, James, 'Australian Naval Policy 1939-45' in David Steven, Ed., *The Royal Australian Navy in World War II*, Allen & Unwin, Sydney, 1996.

Gordon, A. H., 'HMAS Sydney in Korea: The Firefly Observer' in T. Frame, J. Goldrick and P. Jones, (eds), *Reflections on the RAN*, Kangaroo Press, Sydney, 1991.

Grove, Eric, *Vanguard to Trident*, Bodley Head, London, 1987.

Grove, Eric, 'British and Australian Naval Policy in the Korean War Era' in T. Frame, J. Goldrick and P. Jones, (eds), *Reflections on the RAN*, Kangaroo Press, Sydney, 1991.

Hallion, Richard P., *The Naval Air War in Korea*, Nautical & Aviation, Baltimore, 1986.

Harrison, W., *Fairey Firefly*, Airlife, Shrewsbury, 1992.

Hobbs, David, *Aircraft Carriers of the Royal and Commonwealth Navies*, Greenhill, London, 1996.

Hooker, John, *Korea, the Forgotten War*, Time-Life, Sydney, 1989.

Hopton, John, 'HMAS Sydney Goes to War', *AHSA Journal*, Vol. 21, No. 2, 1981.

Howard, Grant, *The Navy in New Zealand*, Reed, Wellington, 1981.

Hunt, H., 'The Firefly Family', *Aeroplane Monthly*, May 1992.

Hyslop, Robert, *'Aye Aye' Minister*, AGPS, Canberra, 1990.

James, E. T., 'The Career of HMAS Sydney', in *Navy Quarterly*, Vol. 2, No. 1, 1973.

Jones, R. M., 'Fleet Air Arm goes into the Seventies', *Navy Quarterly*, Vol. 2, No. 4, 1974.

Lane, Fred T. and Lane, Gerry, 'HMAS Sydney in Korea: The Sea Fury Pilot', in T. Frame, J. Goldrick and P. Jones, (eds), *Reflections on the RAN*, Kangaroo Press, Sydney, 1991.

Law, Phillip, *The Antarctic Voyage of HMAS Wyatt Earp*, Allen & Unwin, 1995.

Lee, Norman, 'HMAS Sydney in Korea: The Firefly Pilot', in T. Frame, J. Goldrick and P. Jones, (eds), *Reflections on the RAN*, Kangaroo Press, Sydney, 1991.

Lynch, Thomas G., 'Warrior and Magnificent', in John Roberts, Ed., *Warship 1995*, Conway, London, 1995.

Mackay, Ron, *Hawker Sea Fury*, Squadron/Signal, Carrolltown, 1991.

MacPherson, Ken and Burgess, John, *The Ships of Canada's Naval Forces 1910-1981*, Collins, Toronto, 1981.

Marriott, Leo, *Royal Navy Aircraft Carriers 1945-1990*, Ian Allan, London, 1985.

McDougall, R. J., *New Zealand Naval Vessels*, GP Books, Wellington, 1989.

McGuire, Frances, *The Royal Australian Navy*, Oxford University Press, Melbourne, 1948.

McMurtrie, Francis E. and Raymond Blackman, V. B., *Jane's Fighting Ships*, Sampson Low, London, annually.

Meyers, Edward C., *Thunder in the Morning Calm*, Vanwell, St Catharines, 1992.

Morison, Samuel Eliot, *Victory in the Pacific*, Little, Brown, Boston, 1960.

Murfett, Malcolm, *Hostage on the Yangtse*, USNI, Annapolis, 1991.

Murfett, Malcolm, *In Jeopardy*, Oxford University Press, Kuala Lumpur, 1995.

Murray, Rear Admiral D. S., 'The Sea War in Korea 1950-1953' in *Naval Historical Review*, Vol.1, June 1976.

Nesdale, Iris, *Action Stations!*, HMAS Warramunga Assoc., Adelaide, 1989.

Norton, Frank, *Fighting Ships of Australia and New Zealand*, Angus & Robertson, Sydney, 1953.

Oliver, David, *British Combat Aircraft in Action since 1945*, Ian Allan, London, 1987.

O'Neill, Robert, *Australia in the Korean War 1950-53, Vol. 2*, Australian War Memorial & AGPS, Canberra, 1985.

Palmer, Michael, 'The U.S. Navy and the Persian Gulf', in William Roberts and Jack Sweetman, (eds), *New Interpretations in Naval History*, USNI, Annapolis, 1991.

Parker, R. G., *Cockatoo Island*, Nelson, Melbourne, 1977.

Payne, M. A., *HMAS Australia 1928-1955*, NHSA, Sydney, c1975.

Polmar, Norman, *Aircraft Carriers*, Macdonald, London, 1969.

Rawlings, John, *Pictorial History of the Fleet Air Arm*, Ian Allan, London, 1974.

Richmond, H. W., *Statesmen and Sea Power*, Oxford University Press, London, 1946.

Roberts, W. O. C., 'Gun Battle on the Han', *Naval Historical Review*, Vol.1, No.2, September 1976.

Roy-Chaudhury, Rahul, *Sea Power and Indian Security*, Brassey's, London, 1995.

Scott, J. D., *Vickers: A History*, Weidenfeld & Nicolson, London, 1962.

A Few Memories of Sir Victor Smith, Australian Naval Institute, Canberra, 1992.

Sinclair, Keith, *A History of New Zealand*, Penguin, Auckland, 1980.

Sinclair, Keith, (ed.), *Tasman Relations*, Auckland University Press, 1987.

Soward, Stuart, 'Canadian Naval Aviation, 1915-69', in James Boutilier, Ed., *RCN in Retrospect*, UBC, Vancouver, 1982.

Sturtivant, Ray, *British Naval Aviation*, Arms & Armour, London, 1990.

Swanson, Bruce, *Eighth Voyage of the Dragon*, USNI, Annapolis, 1982.

Thetford, Owen, *British Naval Aircraft Since 1912*, Putnam, London, 1978.

Walker, Martin, *The Cold War*, Fourth Estate, London, 1993.

Watson, Bruce, *The Changing Face of the World's Navies, 1945 to the Present*, Arms & Armour, London, 1991.

Watts, A., *A Source Book of Aircraft Carriers and their Aircraft*, Ward Lock, London, 1977.

Weaver, Trevor, *Q Class Destroyers and Frigates of the Royal Australian Navy*, NHSA, Sydney, 1994.

Whelan, Richard, *Drawing the Line*, Little Brown, Boston, 1990.

Williams, Ray, *Fly Navy*, Airlife, 1989.

Willoughby, C. A. and Chamberlain, J., *MacArthur 1941-1951*, McGraw-Hill, New York, 1954.

Wilson, Stewart, *Sea Fury, Firefly and Sea Venom in Australian Service*, Aerospace, Canberra, 1993.

Winton, John, *Air Power at Sea*, Sidgwick & Jackson, London, 1987.

Zammit, Alan, 'I'm Jesus, the Canteen Manager', a series of six articles in *Naval Historical Review*, Vol. 3, No. 2 to Vol. 4, No. 3, 1981 and 1982.

Ziegler, Oswald, (ed.), *Jubilee*, Ziegler, Sydney, 1951.

PERSONAL RECOLLECTIONS

John Champion   Don McLeod
Ron Christie   David Oliver
Neville Jenkins Alan Porter
Noel Knappstein Charles Thwaites

Chapter Notes

CHAPTER 1
BUILDING A NEW NAVY

1. Mr Dedman's ministerial statement is in *Commonwealth Debates*, 4 June 1947, pp. 3335-3346, *Defence - Post-war Policy*.
2. *Commonwealth Debates*, 4 June 1947, p. 3337
3. *Commonwealth Debates*, 4 June 1947, p. 3338
4. *Commonwealth Debates*, 4 June 1947, p. 3339
5. *Commonwealth Debates*, 4 June 1947, p. 3339
6. *Commonwealth Debates*, 4 June 1947, p. 3339
7. *Commonwealth Debates*, 4 June 1947, p. 3343
8. *Commonwealth Debates*, 4 June 1947, p. 3345
9. *Commonwealth Debates*, 4 June 1947, p. 3336
10. *Commonwealth Debates*, 23 Sept. 1948, p. 834
11. The full quote is in *Commonwealth Debates*, 4 June 1947, p. 3338.
12. Mike Crosley, *Up in Harm's Way*, p. 78
13. Eric Brown, *Wings of the Navy*, p. 156
14. Telegram 30 January 1948 quoted by Eric Grove, 'Policy in the Korean War Era', in T. Frame et al, *Reflections on the RAN*, p. 255

CHAPTER 2
SHOWING THE FLAG

1. *The Age*, 12 July 1947
2. *The Age*, 17 July 1947
3. *Facts and Figures*, June 1945, p. 29
4. *The Age*, 14 July 1947
5. *Sydney Morning Herald*, 23 July 1947
6. *Sydney Morning Herald*, 24 July 1947
7. *The Age*, 15 July 1947
8. *Courier Mail*, 11 August 1947
9. *New Zealand Herald*, 2 September 1947
10. *New Zealand Herald*, 8 September 1947

CHAPTER 3
THE END OF THE BEGINNING

1. *Sydney Morning Herald*, 4 February 1948
2. *Sydney Morning Herald*, 3 February 1948
3. J. D. Scott, *Vickers, A History*, p. 314
4. *The Argus*, 19 May 1949
5. *Port of Melbourne Quarterly*, July-Sept 49
6. *Port of Melbourne Quarterly*, July-Sept 49
7. *The Age*, 19 May 1949
8. *Sydney Morning Herald*, 24 May 1949
9. *Commonwealth Debates*, 29 Sept. 1948, p. 959
10. *Sydney Morning Herald*, 4 June 1949

CHAPTER 4
HIS MAJESTY'S AUSTRALIAN FLEET

1. An official statement is in *Commonwealth Debates*, 19 September 1947, p. 95-6
2. Norman Friedman, *The Postwar Naval Revolution*, p. 225.
3. *Sydney Morning Herald*, 5 March 1952
4. Robert Hyslop, *Aye Aye, Minister*, p. 72
5. *Sydney Morning Herald*, 9 August 1949
6. *Courier Mail*, 30 September 1949
7. *Commonwealth Debates*, 28 September 1949, p. 710
8. *Courier Mail*, 30 September 1949
9. *Commonwealth Yearbook*, 1954, p. 1106

CHAPTER 5
ANZAC BROTHERS

1. *The Dominion*, 27 February 1950
2. *The Dominion*, 6 March 1950
3. *The Dominion*, 6 March 1950
4. *New Zealand Herald*, 20 March 1950
5. *New Zealand Herald*, 27 March 1950
6. *Sydney Morning Herald*, 8 May 1950

CHAPTER 6
THE SHADOW OF THE NEW WAR

1. *Sydney Morning Herald*, 5 May 1950
2. Eric Harrison, Australia's Defence Plans, in Oswald Ziegler, (ed.), *Jubilee*, p. 253
3. Keith Sinclair, *A History of New Zealand*, p. 287
4. *Facts and Figures* No.29, p. 6
5. *Commonwealth Debates*, 7 March 1951, p. 78, *Prime Ministers' Conference, London, 1951*
6. *Commonwealth Debates*, 14 March 1951, p. 461
7. *The Age*, 5 June 1947
8. *Commonwealth Debates*, 28 January 1950, p. 3171, *International Affairs*

9 J. D. B. Miller, 'Australasia and the World Outside' in Keith Sinclair, (ed.), *Tasman Relations*, p. 88
10 Quoted in the *Herald*, 25 October 1952

CHAPTER 7
THE LIFE OF THE SHIP

1 Recruiting advertisement in *Sydney Morning Herald*, 26 October 1950.
2 Interview in *Courier Mail*, 9 August 1947.
3 Television interview aboard USS *Independence*, 1995
4 A. H. Gordon, 'The Firefly Observer', in T. Frame et al, *Reflections on the Royal Australian Navy*, p. 292
5 Norman Lee, 'The Firefly Pilot', in T. Frame et al, *Reflections on the Royal Australian Navy*, p. 288
6 *The Age* and *The Argus*, 18 May 1949
7 *The Age*, 17 May 1949
8 Stewart Wilson, *Sea Fury, Firefly & Sea Venom in Australian Service*, p. 85

CHAPTER 8
THE TEST OF BATTLE

1 James Cable, *Navies in Violent Peace*, p. 18
2 *Facts and Figures* No.30, p. 27
3 *Sydney Morning Herald*, 24 October 1951
4 Mike Crosley, *Up in Harm's Way*, p. 114
5 Details of sorties and this report are from secret debriefing reports reproduced in Stewart Wilson, *Sea Fury, Firefly and Sea Venom in Australian Service*, pp. 92-99.
6 From the ship's *War Diary — Korean Operations*, and frequently quoted.
7 *The Age*, 3 March 1952
8 *The Age*, 3 March 1952
9 Norman Bartlett, *With the Australians in Korea*, p. 249
10 W. Roberts, 'Gun Battle on the Han', in *Naval Historical Review*, Sept. 1976, p. 11
11 S. Morison, *Victory in the Pacific*, p. 149
12 *The Age*, 3 March 1952
13 John Bovey, 'The Destroyers' War in Korea, 1952-53', in A. Boutelier, (ed.), *RCN in Retrospect*, p. 251
14 Stewart Wilson, *Sea Fury, Firefly and Sea Venom*, p. 82

15 Letter to his father dated 16 September 1952, in *Up in Harm's Way*, p. 112
16 W. B. Cornish, *Monsoon Lands, Part 2*, p. 122
17 Interview with Vic Cassels, in *For those in peril*, p. 222
18 W. Harrison, *Fairey Firefly*, p. 121
19 John Winton, *Air Power at Sea*, p. 42
20 *The Age*, 3 March 1952
21 *Sydney Morning Herald*, 24 October 1951
22 Vice-Admiral C.T. Joy, quoted in Cagle & Manson, *The Sea War in Korea*, p. 493
23 *The Age*, 30 November 1952
24 Cdr M. U. Beebe of USS *Essex*, interview quoted in Cagle & Manson, *The Sea War in Korea*, pp. 253-4.

CHAPTER 9
THE WAY FORWARD

1 *Sydney Morning Herald*, 7 March 1952
2 Norman Bartlett, *With the Australians in Korea*, p. 250
3 *The Age*, 3 March 1952
4 *Facts and Figures*, No.33, p. 25
5 *Commonwealth Yearbook*, 1953, p. 1248
6 *Commonwealth Debates*, 4 Sept. 1952, p. 990
7 *Commonwealth Debates*, 4 Sept. 1952, pp. 992-3
8 Robert O'Neill, *Australia in the Korean War*, Vol. 2, p. 498
9 Alan Zammit, 'I'm Jesus, the Canteen Manager', *Naval Historical Review* August 1982, p. 31
10 *The Age*, 31 October 1952
11 *The Age*, 24 March 1953
12 Cyril Fields, 'The Fleets of Two Naval Reviews', *The Illustrated London News*, 20 June 1953, p. 1046
13 Arthur Bryant in *The Illustrated London News*, 20 June 1953, p. 1022
14 From the song 'Hughie', in the Fleet Air Arm Song Book, quoted in P. Beaver, *The British Aircraft Carrier*, p. 114. A 'fish-head' is a person on general service sea time.

AFTERWORD

1 C. A. Willoughby & J. Chamberlain, *MacArthur*, p. 297
2 *The Age*, 1 May 1954

INDEX

GENERAL

Adelaide 17, 18, 58
Akaroa 48, 49
Amberley Air Base 17, 23, 27
Anstice, Edmund 21
ANZUS Treaty 60, 61
Appleby, John 75
Arbroath 30
Armstrong, John 21
Atomic Bomb 42, 53, 98, 100
Auckland 23, 24, 27, 45, 49, 50, 58, 60, 104

Babbit, Arlene 84
Bailey, 'Bugs' 86, 89
Barnett, R. 55
Barrow in Furness 29
Bathurst Island 105
Battle of the Islands 87
Bay of Islands 49
Beecher, Commander 32
Beecroft Head 55, 71
Berlin 28, 52
Bidwell, R 104
Blanfield, A 45
Bougainville Reef 41
Bowles, Jim 75, 100
Bradley's Head 33, 75
Brisbane 23, 24, 35, 41, 58, 74, 98
British Pacific Fleet 10, 11, 15, 17, 19, 23, 38, 46, 50, 76
Buchanan, Danny 31
Buchanan, Herbert 102
Burrell, Henry 47, 105

Cairns 35, 96
Canberra 51, 97
Cape Wrath 44

Captain Cook Dock 10
Casey, Richard 38, 60
Chaeryong 83, 89
Chang-San-Got Peninsula 85
Changyen 89
Chifley, Ben 9, 42
China 28, 31, 34, 52
Chinnampo 73, 82
Chonsan 89
Christchurch 49, 104
Clarkson, Keith 84
Clifford Island 75
Cockatoo Island Dockyard 16, 94, 95
Cockburn Reef 35
Coleman, Ron 87, 88
Colquohoun, K. 76
Collins, John 33, 55, 61, 97, 98, 101, 102, 107
Conscription 45, 54, 97
Cooper, Peter 87
Coronation Fleet Review 102
Couchman, W. 20
Crabb, Lt Com 62
Creasy, George 18, 20, 22, 24, 103
Crofts, D. 41
Crosley, Mike 15, 77, 88

Darwin 96, 105
Dedman, John 9, 10, 12
Devonport (NZ) 49
Devonport (UK) 29–31
Dickson, Robert 18, 19
Dollard, Allen 83, 94
Dolphin, G. 102
Dovers, Lt Com 62
Dowling, Roy 28, 30–2, 101
Downard, Commander 62
Dreger Harbour 11, 37
Dunedin 27

Dunlop, Sub Lt 89
Dyer, Admiral 83

Eaton, Rear Admiral 98, 101
Eccles, John 46–8, 56, 103
Eglinton 29
Exmouth Gulf 98
Expenditure 9, 10, 13, 16, 29, 42

Farncomb, Harold 27, 38, 41, 46
Fell, Mike 75, 81
Fitzpatrick, Harry 31
Flinders Naval Depot 47, 57
Francis, Josiah 51, 59
Fremantle 31, 40, 94, 99, 100, 102

Gabo Island 16, 58
Gage Roads 100
Garden Island Dockyard 10, 11, 33, 39, 40, 74, 95, 97, 104
Geelong 47
Gibraltar 38, 103
Glenelg Beach 37
Goldrick, Peter 74
Gooding, Callis 83, 84
Gosling 'Nutty' 19
Gowles, Walther 87
Greymouth Coal Mines 57
Guadalcanal 25, 98

Haeju 54, 75, 83
Halsey, William 107
Hamilton, Louis 13, 38
Hancock, Valston 11
Hancox, Phillip 83
Hanna, Pat 44
Han River 77, 82, 83, 85, 94
Harries, David 44, 75, 79–81, 91–2, 94, 95
Harris, John 27
Hauraki Gulf 25, 49
Heard Island 28, 39, 40
Herrick, L. 49
Hervey Bay 74, 105
Heyward, Allan 66

Hobart 18, 24, 27, 35, 40, 47, 56, 57, 104
Holland, Sidney 58
Hong Kong 27, 31–2, 45, 52, 60, 92, 107
Hungnam 85

Inchon 53, 54, 90
Iwakuni Air Base 82

Jamestown Line 82
Japan –Occupation 8, 24, 36, 37, 41, 53
 –Peace Treaty 60
Jervis Bay 11, 22, 32, 35, 44, 46, 47, 55, 71, 74, 104
Johnston, D. Hammersley 47

Kansong 85
Keene, J. Ruck 56
Kempsey 40
Kerguelen Island 28
Kiama 22
Kimpo 84
Kingston, water quality 104
Knappstein, Noel 28, 66, 84
Kojo 77
Kure 37, 87, 88

Lascelles, O. 49
Lee, Norman 67, 86, 89
Litchfield, Geoff 67
Lord Howe Island 40
Lossiemouth 29
Lowndes, Peter 40
Lunberg, Richard 75, 84, 88–9

MacArthur, Douglas 60, 107
Macdonald, Thomas 49
MacMillan, Neil 83
Macquarie Island 28, 39
Malaya 28, 52, 60, 97, 102
Manpower 11, 12, 25, 33, 44, 45, 54, 63, 96, 97
Manus Island 11, 24, 37–9, 42, 59, 96–9, 105, 108
Mao Zedong 52, 80
Martin, Harold 26, 27, 91

INDEX

McKell, William 94
McMahon, William 96, 97
McPherson, Ken 24
Melbourne 16, 18, 19, 29, 32, 35, 39, 40, 47, 51, 94, 95, 97, 100–2
Melbourne Cup 35, 47, 84, 101
Menzies, Robert 54, 58–60, 98, 102
Milford Haven 30
Monte Bello Islands 98–100
Montgomery, Field Marshal 22
Mountbatten, Louis 103
Murray, Brian 19

Nankervis, Alfred 21
Napier 49
NATO 60, 73
Newcastle 40, 95
New Guinea 8, 24, 35–7, 41
New Plymouth 57
Nicholls, David 59
Nowra 11, 22, 29, 32, 55, 74, 75, 77, 94, 100, 102

Oakley, Albert 87, 102
O'Donovan, Frank 88
Oldham, George 47, 62, 105
Oliver, Dave 63, 67
Ongjin 89
Operation Strangle 74, 89

Pago Pago 38
Pearl Harbor 26, 104, 108
Pedder, A. 95
Peek, Richard 90, 94
Pick, SCO 62
Pohang 54
Point Cook 18, 20, 21, 35, 47
Poole, Robert 32
Port Arthur (China) 80
Port Arthur (Tas) 57
Porter, Alan 28, 31, 41
Port Phillip 18, 20, 21, 35, 47, 101
Portsmouth 43, 44, 103
Prague 28, 52
Pusan 53

Rabaul 17, 20, 59
Radford, Arthur 50, 61
Rathmines 22, 27
Richmond Air Base 27, 102
Richmond, Herbert 14
Riordan, William 26, 33, 40, 43, 58, 96
Rowell, Sydney 20
Rowland, Sub Lt 89
Russell, Guy 79

Sasebo 80, 81
Schofields 11, 102
Scott-Moncrieff, A. 77, 85, 91, 95
SEATO 107
Segori 89
Seoul 77
Shanghai 32
Shark Island 74
Sicily 45
Simonoseki Strait 37
Simpkin, M. 82
Sinclair, Dick 87
Singapore 20, 24, 39, 43, 92, 107, 108
Smith, Raymond 87
Smith, Robert 55
Smith, Victor 94
Sok To 84, 87
Sokton-Ni 84
Songjin 75, 86
Sortie Rates 73, 75, 76, 79, 80, 88, 90
Spender, Percy 60
Stalin, Joseph 80, 93
Station Pier 18, 19, 21, 32, 101, 102
St Merryn 43
Storm Bay 56, 57
Subic Bay 39
Suez Canal 12, 55, 102
Sydney 11, 16, 22, 23, 26, 27, 35–8, 43, 44, 46, 48, 51, 56, 58–60, 70, 75, 92, 94, 97, 99–102

Taewha Do 82, 87
Tasman Sea 47, 55
Thomson, Lt 62

Townsville 8, 98
Turner, Lt Com 83
Typhoon Ruth 81

Vian, Philip 36
Vietnam 12, 28, 89, 93, 102, 107
Visitors 19, 23, 24, 32, 48, 49, 101
Vladivostok 80, 93

Wackett, Ellis 20
Waiouru 50
War Criminals 20, 38, 59
Webster, Ian 100
Wellington 24, 47, 48, 57
Western Port 27, 35
Whangarei 50
Whenuapai. Air Base 24
Williamtown Air Base 27, 98
Willoughby, Guy 28
Wonsan 85
Wonto 84

Yalu River 82, 84, 87, 90
Yangtse River 25, 31, 37
Yokosuka Naval Base 60
Yonan 89

Zammit, Vic 70

AIRCRAFT

Avenger 27, 92
Barracuda 19
Bearcat 27
Beaufighter 22, 27, 35
Beaufort 46
Catalina 22, 27
Corsair 73, 75
Dakota (DC3) 52, 59
Dauntless 12
Dragonfly (HO3S) 81, 99
Firefly 12, 15, 16, 18–24, 28, 29, 30–1, 40, 43, 44, 46, 48, 55, 56, 66–8, 75, 76, 78–84, 86, 89, 90, 92, 99, 100, 102, 103, 111
Gannet 16, 101

Harvard 50
Hellcat 93
Helldiver 27, 93
Kingfisher 28
Lincoln 16, 22, 27, 52
Meteor 83
MiG-15 80–1
Mosquito 22, 47, 49, 50
Mustang 19, 22, 23, 27, 32, 35, 47, 53, 111
Panther 73
Seafire 16, 18–22, 24, 28, 41, 74, 111
Sea Fury 15, 29, 40, 42, 46, 48–50, 55, 57, 66, 67, 71, 71, 74–6, 78–81, 83, 84, 88–90, 92, 99, 100, 102, 103, 111
Sea Hawk 16
Sea Otter 15, 19, 24, 81, 99, 101, 111
Sea Venom 16
Skyraider 73
Spitfire 28, 31, 92
Sunderland 49
Superfortress (B29) 53, 82
Sycamore 101
Tiger Moth 19, 24–5, 28
Vampire 32, 92
Walrus 28
Wirraway 28, 75
Wyvern 16
Yak-3 15
Yak-9 74

SHIPS

ARGENTINA
25 de Mayo 20

AUSTRALIA
Anzac 9, 59, 74, 75, 98, 105, 107, 108, 110, 112
Arunta 22, 23, 104, 105, 107, 110, 112
Australia 8, 22–4, 26, 27, 31, 35, 37, 40, 41, 46–50, 56–9, 62–4, 95, 98, 101, 104, 106, 109, 112

Barcoo 37, 105, 110
Bataan 22, 24, 26, 27, 35, 41, 46–8, 53, 54, 73–5, 91, 92, 95, 98, 104, 106, 107, 110, 112
Canberra 98
Cerberus 100
Colac 101
Condamine 56, 105, 110, 112
Cootamundra 101
Cowra 101
Culgoa 26, 37, 56, 99–101, 105, 110, 112
Eagle (tug) 40
Gascoyne 7
Gladstone 35, 57, 101
Hawkesbury 99, 105
HDML 1328 39
Hobart 8, 24, 30, 95, 109, 112
Jabiru 37
Kangaroo 35, 74
Kanimbla 8, 29, 110, 112
Karangi 98
Kimbla 95
Koala 59, 99
Kuramia 106
Labuan (LST 3501) 28, 39, 99, 110
Lachlan 37, 50, 110
Latrobe 57, 101
Limicola 99
Macquarie 59, 99, 101, 104, 105, 110
Manoora 8, 112
Melbourne 29, 34, 51, 101, 102, 107, 108, 109
Quadrant 98, 101, 105, 108, 110
Queenborough 108
Quiberon 26
Quickmatch 26
Reserve 37, 56
Shoalhaven 22, 26, 31, 46, 53, 56, 57
Shropshire 8
Swan 35
Sydney (1913) 7, 33
Sydney (1935) 28

Sydney (1949) 8, 16, 29–33, 40–4, 46–51, 55–9, 62–4, 66, 68–70, 74–92, 94, 95, 99–106, 108, 109, 112
Tarakan 37, 39
Tide Austral 95
Tobruk 9, 16, 29, 55, 56, 58, 59, 75, 82, 85, 86, 90, 94, 95, 98, 99, 101, 104, 105, 108, 110, 112
Vengeance 101, 102, 104–7, 109, 112
Voyager 94
Wagga 95, 101, 106
Warramunga 22, 31, 37, 39–41, 46, 48, 49, 54, 73, 92, 95, 107, 110, 112
Warreen 99
Warrego 37, 98, 105
Warrnambool 35
Wyatt Earp 28

BRAZIL
Almirante Barroso 103

CANADA
Athabaskan 54, 73, 84
Cayuga 54, 73, 78
Crusader 86
Magnificent 16, 68, 92, 103, 104, 109
Ontario 58, 103
Quebec 103
Sioux 54, 73, 82, 85
Warrior 16, 109

FRANCE
Arromanches 93
Dixmude 12
Emile Bertin 12
Lafayette 93
Montcalm 103

GERMANY
Emden 7
Tirpitz 15

INDIA
Delhi 103
Vikrant 21

Netherlands
Banckert 56
Tromp 103
Van Galen 85

New Zealand
Arbutus 24
Bellona 24, 27, 45–50, 56–8, 63, 109
Black Prince 102–5, 109, 112
Echuca 97
Hawea 57, 83, 110, 112
Inverell 97
Kaniere 46, 110, 112
Kiama 97
Lachlan 50, 57, 63, 110
Pukaki 46, 48, 49, 54, 110, 112
Rotoiti 45, 46, 48–50, 83, 110, 112
Stawell 97
Taupo 46, 48, 49, 56, 57, 83, 110, 112
Tutira 46, 48, 54, 110, 112

Pakistan
Shamsher 56
Sind 56

Spain
Miguel de Cervantes 103

Sweden
Gota Lejon 103

Thailand
Prasae 73

United Kingdom
Aeneas 26, 27
Amethyst 25, 31, 34, 84
Amphion 23
Apollo 103
Attacker 41
Belfast 54, 77, 78, 84, 85, 109
Brown Ranger 77
Campania 99–100
Cardigan Bay 98
Ceylon 77
Cleopatra 103
Cockade 18, 20, 21–4, 110
Colossus 15
Comus 78
Concord 78
Constance 85, 87
Contest 18, 21, 24, 30, 110
Cossack 52, 78
Dasher 82
Decoy 103
Defender 103
Devonshire (cruiser) 103
Devonshire (troopship) 94
Diamond 103
Dido 103
Duchess 103
Eagle 103
Gambia 103
Glasgow 103
Gothic (Royal yacht) 37, 94, 106
Glory 13, 17–25, 29, 44, 59, 75–7, 79, 80, 90–2, 109
Illustrious 28, 103
Implacable 44, 103
Indefatigable 103
Indomitable 46, 103
Manxman 103
Narvik 99
Ocean 13, 15, 31, 79, 88, 105, 109
Perseus 103
Plym 99, 100
Rapid 44
Sheffield 103
Superb 103
Surprise 103
Sussex 18
Swiftsure 103
Tactician 43, 56, 104, 110
Telemachus 43, 47–9, 56, 110
Theseus 17–25, 74, 103, 109
Thorough 43, 99, 104, 110
Tracker 99
Triumph 31, 52, 54
Unicorn 31, 81, 82, 87, 92, 110
Vanguard 37
Vengeance 19, 44, 101, 109

Victoria and Albert 103
Vindex 21
Warrior 29, 93, 109
Wave Chief 77
Wave King 99
Wave Premier 77
Whitesand Bay 52
Wilton 44
Zeebrugge 99

UNITED STATES
Antietam 23
Baltimore 103
Bataan 92
Bon Homme Richard 77
Boxer 77
Colahan 78
Collett 54
Essex 77
Forrestal 93
Forrest Royal 73
Hanson 78
Helena 77
Higbee 54
Hyman 85
Independence 80
James E. Kyes 54
Juneau 54
Keppler 26
Lloyd Thomas 26
Los Angeles 77
LSMR 401 85
LSMR 403 85
LSMR 404 85
Mispillion 26
New Jersey 77–9, 85, 109
Oriskany 81
Princeton 80
Rendova 75, 77
Renshaw 78
Saratoga 93
Shangri La 23
Shields 78
Sicily 77
Tarawa 107
Toledo 77
Valley Forge 26–8, 73, 74, 109
William C Lawe 26
William M. Wood 26

USSR
Sverdlov 103